POINTS OF VIEW

POINTS OF VIEW

A SERIES OF BROADCAST ADDRESSES

By

G. LOWES DICKINSON / DEAN INGE

H. G. WELLS / J. B. S. HALDANE

Sir OLIVER LODGE / Sir WALFORD DAVIES

With
AN INTRODUCTION AND
A SUMMING-UP BY

G. LOWES DICKINSON

and

A SUPPLEMENTARY LETTER BY

Sir OLIVER LODGE

LONDON

GEORGE ALLEN & UNWIN LTD

MUSEUM STREET

First published, with the exception of the Introduction, in serial form in "The Listener"

FIRST PUBLISHED IN BOOK FORM 1930

PRINTED IN GREAT BRITAIN BY
UNWIN BROTHERS LTD. WOKING

CONTENTS

PAGE

INTRODUCTION, BY G. LOWES DICKINSON 9

I. G. LOWES DICKINSON 19

II. DEAN INGE 33

III. H. G. WELLS 49

IV. J. B. S. HALDANE 71

V. SIR OLIVER LODGE 91

VI. SIR WALFORD DAVIES 111

VII. A SUMMING-UP, BY G. LOWES DICKINSON 127

VIII. SUPPLEMENTARY LETTER BY SIR OLIVER LODGE 145

INTRODUCTION

THE addresses here reproduced were delivered under the auspices of the British Broadcasting Corporation and are a small instalment of an experiment in popular education. Such methods will not, of course, supersede schools and colleges, still less books, but they may reach a larger and more miscellaneous audience. They are seeds thrown out over the world at large, and their crop will help to determine its future. Public opinion is now in the making, not only in the West but in the East, and this, along with many others, is one of the ways in which the great question is being put and answered—Can democracy succeed?

If such talks are to serve their purpose, certain snares must be avoided. First their educational purpose must be maintained. There are countries where broadcasting is prostituted to advertisement. That is to make the temple a den of thieves. Let us hope that the Corporation will continue to show, what it has shown up to now, the will and the power to resist such invasions. Next, every important point of view should have a fair hearing. The present series has been criticised as leaning too much to the side of free thought and too little to that of orthodox religion or conservative politics. The criticism, for what it is worth, is just, though the choice of speakers was not made with that object, and one of them is a high dignitary of the Church of England. But ours is only one specimen of the work the B.B.C. is continually performing, and no one will suggest that it is typical. The Churches and the conservatives, cannot be said, on the whole, to suffer from under-representation.

What I have to say here by way of introduction does not in any way commit the other speakers, nor, I hope, will it offend readers who may be altogether out of sympathy with it. These very likely will be found both on the left and on the right, and I shall be charged both with dull platitude and revolutionary provocation. That is all as it should be, for so opinion shapes itself; I will only ask the reader to believe, as no doubt he will, that I, like the rest of the speakers, am serious and sincere. The topics treated in these talks are very numerous, and I shall make no attempt to deal with them all. But, broadly, two fundamental questions keep recurring—How is truth arrived at? and what is now true about the condition of our society? I shall briefly state my own attitude to both issues, referring, as occasion demands, to the concurrent or diverse views of my colleagues. And first, as to the nature and criterion of truth.

One of the greatest changes that has taken place during the last few centuries in the West is the break-up of the coherent system of the Christian Church. Christianity was, in mediæval times, not only an authoritative doctrine but a political power, co-ordinate with if not superior to the State, condemning heresy and handing the heretic for punishment to the civil power, having the monopoly of education and occupying the highest offices in the State. That imposing, if always threatened, unity was destroyed at the Reformation. The Church of Rome indeed still rears its front of immutable and absolute dogma. But it cannot impose its rule by force even in Catholic countries, and elsewhere it is merely one of the sects. That is one great historical change, and there is another even greater. The East, with its religions, has invaded the spiritual horizon of the West. No longer is it thought possible or desirable—as it was at the time of the Crusades—to convert by force the non-Christian world. On the contrary, the East is making its conquests in the West. Religion is no longer a crusted loaf, it is a ferment in the dough, and it would be rash for any man to say what bread may come out of it. We may approve or disapprove these facts, but we cannot deny them. They determine all our present-day attitudes.

But further, another transformation, even more radical, has occurred. The last few centuries have seen the growth of physical science and the extension of its methods to the whole field of human experience. Of the material consequences of this revolution I shall not here speak. They are a succession of daily miracles. My point is that the method has come, and come to stay. And the essence of the method is that matters of fact, as distinguished from "values", can only be ascertained by evidence based on the senses and controlled by observation, experiment, and inference. Now all the great religions have professed to know, by revelation, some of the most important matters of fact. They have affirmed, for instance, that there are individual souls that survive death; that bodies decayed and scattered long ago re-collect their elements and appear for punishment or reward; that there is a heaven and a hell and a purgatory, or a long series of reincarnations; above all that there is a God, and that we know His nature, and that He takes active part in our destiny. There is also an account of the origin of evil. When I referred to this account, as given in Genesis, I was reproached with deliberate misrepresentation. I was not intending to misrepresent, nor do I think I did, the belief of enormous numbers of Christians. And the very fact that others can treat the story as a myth shows how enormously our mental attitude has changed. Yet the full implications of the change are not always grasped. They are, that on every question of "fact", as distinguished from moral or æsthetic judgments, there is only one authority, and that is science. In this series, for example, Sir Oliver Lodge tells us that he is convinced of the survival of individuals after death. But he is convinced by evidence which he believes to be sound, and would be open to conviction if the evidence should be successfully challenged. The gulf between this attitude and that of mediæval Christianity is profound. It implies the greatest and most pregnant change that has ever happened in the world.

Now, on those great questions of fact which the religions professed to answer, science has not pronounced; what then, it may be asked, is the plain man to do about them? There is, as there always has been, a very small number of people who profess to know the truth by a special vision commonly called "mystic". The great mystics say that they have experienced this truth directly. But their evidence cannot be tested except by the very few who may have had similar experience, and we cannot, the rest of us, do anything about it, except wonder. It cannot form part of our science, nor, therefore, of our knowledge of fact. On the other hand, it does call attention to the limitations of our normal experience. What we know is what is knowable to senses like ours and minds like ours. We do not know what the world would look like if we were crabs or butterflies or beetles, not to mention the whole range of microscopic creatures. Our whole life and experience is thus shut up in a hole, though we have the curious power of knowing that we are in the hole. Some people seem to draw from this fact the conclusion that we may guess and believe anything we like about what lies outside it. I do not agree. But neither do I think that we are never to look out. As we come to know about the world from our own angle of vision, we build up a foundation to stand upon. If we fling ourselves down from this eminence we are likely to come to grief. But there is no reason why we should not strain our eyes into what is unknown, but possibly knowable. The people who use their eyes for this purpose, and help us to use ours, are, I think, the artists and poets. They do not dogmatise, they symbolise; and they never mistake, nor should we, their symbols for science. They may, however, give science a lead by giving imagination wings. I tried to put this position in my last talk, and I will not dwell further upon it.

To turn now from fact to what philosophers now call "values"—that is, our ideas about Good and Evil—on these science has nothing directly to say, though it is always influencing our values by its discoveries of fact. Still, in the very last resort, values are dogmas. But wise men do not make up their dogmas without much converse with their kind; the more so as the dogmas that are most important are social. In the last resort, however, a man who believes in values which society rejects may have to suffer martyrdom for his pains. The men the world thinks greatest have done that. It is only after their martyrdom that they come to be accepted, and those who accept them are none the less ready to martyr any other man who wants to revive or extend their discoveries. Values, therefore, at the extreme edge, are solitary and perilous. They build up, nevertheless, in the long run, the growing consent of men about Good and Evil.

These are my heresies, commonplaces, or whatever they may be, about science and about values. There remains the application of both to society. Something all of us in these talks have said about that, and we do not always agree. We seem, however, all of us, in some sense or other, to believe in modern democracy; or at least not to want to pull it down. For even Dean Inge says that he is not prepared to put anything else in its place. Democracy, of course, is a vague word. But these talks are not treatises on political science, and for my own part I still regard the essence of democracy as government by free opinion. I do not know that any of my colleagues would dispute this. But it is being disputed, in practice if not in theory, over a great part of Europe, and I would not guarantee that it may not be disputed in our own country during the lifetime of some now alive. The reason is, that as opinion becomes conscious and active throughout the whole community, it will demand very great changes in the system of property. I have indicated what I think those changes should be, but my views are, of course, moderation itself compared to some that are held. Yet they have shocked some people. Perhaps, however, that may be because they have also been misunderstood. I notice, for instance, that Dean Inge accuses me of undermining the impulse to work. I do not see how I do this by suggesting that all property should take the form of payment for work. As things are, of course, a great part of our property does not take this form and therefore notoriously encourages enormous idleness. I think this very undesirable. But I am sure that when a serious attempt is made to alter it, an opposition may develop of which in recent years we have had no experience in this country. I am glad to note that both Mr. Haldane and Mr. Wells appear to agree with me substantially on this point of property, and I have only to add, what they would not dispute, that any redistribution of property will have to be accompanied by a great development—such as, in fact, is going on—of the application of science to industry. If that proceeds as it is proceeding, and if, at the same time, social institutions are so far improved that the product does not all accumulate in the hands of a few multimillionaires, we shall be getting nearer to the kind of democracy which I want to see, and which, I believe, most reasonable and instructed men and women want to see. All our hopes, however, may be destroyed if we drop government by consent and try to substitute government by violence. That is what Russia has done, and we can hardly be surprised, if we remember what the government of Russia was, both when we English were fighting it as an embodiment of tyranny, and when we were allied with it as a friend of democracy and liberty. But the Russian experiment, if inevitable, is not, so far as I can see, encouraging in its results, and anything of the same kind in the countries of the West would be suicidal. Communists, however, are not in this, or in any Western country, a menace, and they will not become one if we have the wisdom to do by consent what they hope in vain to do by violence. Mr. Bernard Shaw's *Apple Cart* is not an inspiring spectacle. But our condition would not be improved if we upset the cart and took to scrambling for the apples.

In conclusion, it is my belief that science has put it into our power to achieve, at last, if we have the wisdom and the courage, the age-long dream of brotherhood and peace. At the end of my last talk I cited a poem of Goethe's to indicate the spirit which should inspire good men. I will add here a few lines from the Poet Laureate, which sum up better than I could what now lies within our grasp:

> Now music's prison'd raptur and the drown'd voice of truth
> mantled in light's velocity, over land and sea
> are omnipresent, speaking aloud to every ear,
> into every heart and home their unhinder'd message,

the body and soul of Universal Brotherhood;
whereby war faln from savagery to fratricide,
from a trumpeting vainglory to a crying shame,
stalketh now with blasting curse branded on its brow.[1]

This vision is not certitude. But it is a reasonable hope which we can fulfil, if we have patience and understanding and will.

G. LOWES DICKINSON

[1] *The Testament of Beauty*, Bk. I, line 729.

I

G. LOWES DICKINSON, M.A.

FROM WHO'S WHO (1930)

Dickinson, Goldsworthy Lowes, Fellow King's College, Cambridge; *s.* of late Lowes Dickinson, artist. *Educ.:* Charterhouse; King's College, Cambridge. *Publications:* From King to King; Revolution and Reaction in Modern France; The Development of Parliament in the Nineteenth Century; The Greek View of Life; The Meaning of Good; Letters of John Chinaman; Religion: a criticism and a forecast; A Modern Symposium; Justice and Liberty; Religion and Immortality; Appearances; An Essay on the Civilisation of India, China, and Japan, 1914; The European Anarchy, 1916; The Choice before Us, 1917; The Magic Flute, 1920; War: its Nature, Cause and Cure, 1923; The International Anarchy, 1904–14, 1926; (with F. Melian Stawell) Goethe and Faust; an Interpretation, 1928.

I

G. LOWES DICKINSON

THE series I have been asked to open is called *Points of View.* The speakers are invited to say what they think about things in general; and there should at least be no lack of variety. I cannot see Mr. Bernard Shaw agreeing with Mr. Wells, nor Dean Inge with myself, nor Mr. Haldane with any of us.[1] My only puzzle is why I have been selected to begin. It must be, I think, because I have devoted much of my time to the effort to understand and express other people's points of view. The result has been that I have arrived at views of my own. And I shall now state what they are. Half an hour is a short time in which to express one's attitude to the universe. But no matter. I will "leave my damnable faces" and begin.

I begin with politics, in which most Englishmen are interested. I am a democrat; by which I mean that I believe in free and open discussion, about all laws and institutions that exist or ought to exist, and in the right of anybody and everybody to advocate their change, not by force, but by persuasion. When I was young this was a view very generally accepted. But since the war for democracy was

[1] Mr. Bernard Shaw's talk is not included here, by his own choice. It will be found in *The Listener*, Oct. 23, 1929.

fought it has almost become a paradox. Russia and Italy are governed by minorities, who have seized and hold power by force. Poland, the Balkans, and Spain are in much the same case; and apart from the happy little States that have leisure to pursue civilisation because they do not pursue war, only France, the British Empire, and the United States have even the semblance of democracy.

I say the semblance because, as I think, democracy cannot become a reality until we have made radical changes in our system of property. The enormous inequality betwen rich and poor gives all real power to the rich; and it is they who govern, whoever has the vote. They may be, some of them, as intelligent and as public spirited as most of them claim to be. But government by a rich minority is not democracy. If democracy is to be real, the main source of individual incomes must be work, and pay must be, if not equal, as Mr. Bernard Shaw insists it should be, at any rate very much more equal than it is now. Above all, inequalities should not be perpetuated by inheritance. To own property once meant to be liable to public service, and with the inheritance of the property went the inheritance of the obligation. But this has long ceased to be the case; and that the son or daughter of a millionaire should inherit his millions, with no corresponding obligations to employ them in the public interest, is the sort of thing we should think merely lunatic, if we had not got the habit of thinking that everything that exists is sane.

It is equally absurd that a man can buy up land, here or abroad, for a song, and sit doing nothing until it is worth millions. There is much more that might be said of the same kind, but it need not be said here, for most of you will be familiar with this line of argument, either to accept or to denounce it. I have said enough to show where I stand on this issue. I think that all private property ought to take the form of wages, that is, payment for work. But I include, of course, every kind of work, and especially the kind that is most necessary if wealth is to be increased—namely, organising ability and scientific research. But if wealth is to be increased it must not be wasted, and the principal waste is that caused by war. I do not believe it is possible to have a democratic society, as I am conceiving it, and to have also war. True, such democracies as we have had in the past have made war; but they have always been destroyed by it. For the natural outcome of war is dictatorship, as anyone may see now who looks out upon the world. Moreover, it is impossible, while war continues, for wealth to increase to such a point as will make a real democracy possible. I am putting the case very mildly, for I am myself convinced that war, if it continues, will destroy, not only democracy, but mankind. One reason, indeed, why I am opposed to both Fascism and Bolshevism is that both believe in war, one international, the other civil. These two forms of society may appear to be opposite; but, in fact, they are the same figure, only standing in the one case on its head, in the other on its feet. Which is which the listener may decide for himself. It is the figure itself that I object to, whether it stands on its head or its feet.

War then, in my judgment, is the greatest enemy of democracy, and it is none the less so when it calls itself war *for* democracy. What can we do then about war? I believe, myself, in the League of Nations, and there I come into conflict with Mr. Wells. But our disagreement does not seem to me as important as perhaps it does to him. He thinks the League is a futile piece of camouflage. I think it is a very small baby whose life is rather precarious. But I don't want Mr. Wells, for that reason, to empty it out in the bath water. On the whole, I think it is growing stronger and has a good chance of surviving to become a man, if Mr. Wells, and men less able and pacific than he, will let it alone. And I see no alternative. It's easy to say we ought to have a World State. Perhaps we ought. But the same forces that keep the League of Nations weak make a World State chimerical. The League is actually down there in the arena fighting the wild beasts, and if it does not kill them I do not know what will. But if I disagree with Mr. Wells on this point, I agree with him in my opposition to everything which makes States more self-supporting, and therefore less willing to make concessions and agreements with one another. An imperial tariff, for example, may or may not be economically sound—I will not disucss that; but it will certainly make the British Empire what it has never yet been—an object of hostility to all other States. International agreement about Tariffs and raw materials, is the only road to peace; closed, or even semi-closed economic systems, lead towards war. Here, too, the League of Nations is working on the right lines. It is not indeed an infant Hercules, but at least it is trying to strangle the snakes.

What I want to see then is a number of democratic societies, settling their disputes in peaceable ways and devoting all their political and economic energies to the increase of wealth. But why? If men can escape from the hell they have made of their societies, what would or could or should they then do with their lives?

It is often said or implied by scholars, artists, and others, that democracy is, by its nature, materialistic; and in the present condition of the United States, or of England and the Dominions, there is plenty to support that view. But, in fact, these are rather plutocracies than democracies, in the sense in which I am using that term. I will not, however, insist upon that. For the question raised is deeper.

It is thought that men cannot be comfortable without becoming gross, and that the light of the ideal cannot flash except in a tornado of ruin. If men were really like that, I should myself prefer that they should be comfortable and gross rather than tortured in the mass in order that there might arise a few saints and artists. But I mistrust these general views of human nature. I do not pretend to instruct anyone on that subject. But I will lay before you a few ideas for your consideration.

If you were care-free and had leisure, what would you do with it? Sport? Well, you might do worse. At any rate, that is better than war. Physical adventure and risk? The aeroplane and the motor-car provide plenty, though the risks of the latter seem to apply mainly to non-motorists. Hobbies of every kind? Why not? The pictures? Well, they may become better than they are, and they will, if people want them better. The drama? Not very bright in England; but that, perhaps, rather because of the economic conditions of the theatre than for any other reason. Literature? It still exists, it is even still created, behind the enormous and unceasing bombardment of the Press and the novel —those swarms of locusts that die as soon as they are born. I am making, it will be observed, no highfaluting claim for the ordinary man and woman. But I see no reason to suppose that in a democracy they will be less alert, less vital, and less full of the lust of life than the generations of slaves and serfs that have preceded them in history.

That is all very well, some one may say, but how gross! What about the higher values? What about knowledge, art, literature, religion? Well, to which of these would democracy be less favourable than any other form of society? Let us consider.

So far as knowledge is concerned, there has never been an age, in known history, that has offered either the promise or the achievement of the modern world. I am not speaking now of the practical applications of science, important though these are. I am speaking of knowledge for its own sake. Like most people, I do not myself understand physics, and I never shall. But no one can read the books of Professor Eddington without feeling his imagination profoundly stirred. And what must it be to *be* Professor Eddington, or Professor Jeans, or any of their eminent colleagues? Or take biology. This new view of the process of life in which man, instead of being an exceptional creature, damned because his ancestor ate the fatal fruit, is seen as a phase in a vast hierarchy, starting below and reaching beyond our ken—what an inspiration to anyone who can come out into the open and endure the fresh air of dawn! Or psychology. That infant science, it is true, is playing all the pranks which terrible children are known to play. But who can doubt that out of it is coming a new power to understand and therefore to manage that most difficult of all engineries, ourselves. There is nothing wrong with knowledge, except the use we may make of it.

And that brings me to the next point. The use men are going to make of their knowledge depends upon the quality of their imagination. What is it they are going to desire and to will as good? This really is always, in every society, the crucial question. In the earlier part of this talk I have been saying what I myself desire, and what I believe that large numbers of others desire, in the way of social reorganisation. And I shall now try to say what I want people to desire in these higher regions, and what I think many of them do, and more will, desire.

Let us take then, first, the large region of literature and art. I do not know, and no one knows as yet, what the conditions are that make these things possible. We have, for instance, great literature and art in ancient Greece and in mediæval and renaissance Italy, great painting in modern France, great music in Germany. And we have had practically nothing great in any of the arts in any of the new countries. It is, however, I think clear that the form of government, and even of society, has very little to do with the matter. There is no reason, that I can see, why a modern democracy, as I am conceiving it, should not be favourable to great art. If, for instance, the British people cared enough about it, and had the necessary gifts, their democracy might devote more magnificent resources to architecture and painting than the Athenians did under Pericles, or the Venetians under their Doges, or the French under Louis XIV. It will probably be a different architecture from any we have now, and it may be created by engineers instead of by architects. But it will exist. Nor is there any reason why buildings should not once more be covered with great paintings, as they were for generations in Italy. On the other hand, there is nothing to show that government by oligarchies has, in itself, been favourable to art, though great art has occurred under those conditions. During the nineteenth century, in most countries of the West, rich men have governed and still govern. What have they done for art? Look round and see! Nor will democracy do anything, unless it gets a new Spirit. Perhaps it will. Italy, for example, if ever it digests Fascism into democracy, might recover its ancient genius, and Russia, if it digests Bolshevism into democracy, might develop a new one. Even as it is, a French visitor to Russia told me how he listened in dismay and terror to a violent exhortation at a public meeting, supposing it must be an incitement to new massacres, but discovered on inquiry that the orator was merely explaining the importance of taking Proust as a model for literary style.

Which reminds me that literature, from the nature

of the case, is less capable of direct stimulation by government than the other arts—a fact which is important for us in England, because we have had, perhaps have still, more genius for literature than for any other art. And clearly, I should say, the democracy of which I am thinking need not be less favourable to literature than our present anarchy. It might easily be more so. On the other hand, music and drama might and should be supported by public authority. If both flourish in Germany, as they do nowhere else, it is partly because there the theatre and opera are subsidised by the State; and if both have sunk to the lamentable condition they are in here, it is at least partly because the exigencies of finance drive out everything which has not, evidently or supposedly, capacity to attract money. Big business has got hold of the arts and is throttling them. For it is rare indeed for a financial magnate to have a glimmer of imagination outside business! But I must not say more on this subject. I have wished merely to explain why I do not think that a true democracy need be inimical to literature and art, though certainly I dare not prophesy that it would, of necessity, be favourable to them. Everything will depend on the kind of people we show ourselves to be, if and when we have the power to be what we really want to be.

There remains a point which is, to my mind, of very special importance, and on which I want, therefore, to say a few words. What is the possible position of religion in a democratic and scientific age? I am going to speak frankly on this subject, though, as I hope and intend, without giving offence to those who may think very differently from myself. Religion, in my judgment, is not a fixed body of doctrine revealed by Buddha or Jesus or Mahomet. Still less is it the charge of a particular Church pretending to know matters of fact independently of and in contradiction to science. Nor, on the other hand, is it merely a superstition to be got rid of wholesale as soon as possible. If I might attempt what is perhaps impossible—a definition—I would say that what I mean, when I talk of religion, is an attitude of the passionate and informed imagination towards the whole world, and man's place in it. Such an attitude, if one is intelligent, honest, and bent upon truth, is necessarily being continually modified by new truth. Nor is this new truth necessarily either pleasant or encouraging to that fearful, timid, egotistical creature which most of us are. The tragedy of the world goes very deep, and cheap comforts will never abolish it. But as the imaginary tragedies of poets, if they are great enough, leave the spirit not crushed but strangely purged, so can it be with the tragedy of real life. It is when we are shot out of ourselves, naked and shivering, into the vast world, that the great illuminations come; and art is their interpreter. Thus, to my mind, one of the most religious spirits that ever lived is Mozart, and one of the greatest religious experiences his *Magic Flute*. This may be a feeling personal to myself, but others may feel the same about Beethoven or Bach; though, I confess, I find it hard to believe that anyone has felt it in jazz. No matter. You will understand what I am driving at.

There is a story told of himself by the poet Goethe, how, when he was a boy, he built an altar in his room and kindled his offering as the sun rose, to the god in whom he believed. So I have a picture in my mind of individual souls raising their altars against the sky. The incense goes stealing up carrying something with it, something which words would only mistranslate, unless they be the words of a great poet. But better than any words, I am inclined to think, for those sensitive to music, is music itself. And now, in this great epoch of invention, we are very near the time when men sitting beside their own fireside will hear all together the same great tones, and experience not indeed the same but a kindred religion. Of that religion the content will be real experience, and the form, what genius has made of that.

But my time is up. I can but hope that, whatever else I may have done or failed to do, I have exposed points enough for the succeeding speakers, if they choose, to take up, controvert, or develop.

II

DEAN INGE

FROM WHO'S WHO (1930)

INGE, VERY REV. WILLIAM RALPH, D.D., C.V.O. 1918; F.B.A. 1921; Dean of St. Paul's since 1911; *b.* Crayke, Yorkshire, 6 June, 1860; *e. s.* of late Rev. William Inge, D.D., Provost of Worcester College, Oxford, and Mary, *d.* of Ven. Edward Churton, Archdeacon of Cleveland; *m.* 1905, Mary Catharine, *d.* of late Ven. H. M. Spooner, Archdeacon of Maidstone, and *g.d.* of Bishop Harvey Goodwin; three *s.* one *d.* *Educ.:* Eton: King's College, Cambridge. Bell Scholar and Porson Prizeman, 1880; Porson Scholar, 1881; Craven Scholar and Browne Medallist, 1882; Senior Chancellor's Medallist, 1883; 1st class Classics, 1882 and 1883; Hare Prizeman, 1885; Assistant Master at Eton, 1884–88; Fellow of King's, 1886–88; Fellow and Tutor of Hertford Coll., Oxford, 1889–1904; Select preacher at Oxford, 1893–95, 1903–5, 1920–1921; Cambridge, 1901, 1906, 1910, 1912, 1913, 1920, 1924; Bampton Lecturer, 1899; Hon. D.D., Aberdeen, 1905; Paddock Lecturer, New York, 1906; Lyman Beecher Lecturer, Yale, 1925; Vicar of All Saints', Ennismore Gardens, S.W., 1905–7; Lady Margaret Professor of Divinity and Fellow of Jesus College, Cambridge, 1907–11; Hon. Fellow of Jesus College, Cambridge, of Hertford College, Oxford, and of King's College, Cambridge; Academic Committee Royal Soc. of Literature; Gifford Lecturer, St. Andrews, 1917–18; Romanes and Hibbert Lecturer, 1920; Rede Lecturer, 1922; Hulsean Lecturer, 1926; Hon. D.Litt. Durham, 1920; Sheffield, 1924; Hon. LL.D. Edinburgh, 1923; is a Trustee of the National Portrait Gallery. *Publications:* Society in Rome under the Cæsars, 1886; Eton Latin Grammar (with F. H. Rawlins), 1889, 3rd edition 1900; Christian Mysticism, 1899; two essays in Contentio Veritatis, 1902; Faith and Knowledge, 1904; Selections from the German Mystics, 1904; Studies of English Mystics, 1906; Truth and Falsehood in Religion, 1906; Personal Idealism and Mysticism, 1907; Faith, 1909; Speculum Animæ, 1911; The Church and the Age, 1912; Types of Christian Saintliness, 1915; The Philosophy of Plotinus, 1918; Outspoken Essays, 1919; Second Series, 1922; The Idea of Progress (Romanes Lecture), 1920; The Victorian Age (Rede Lecture), 1922; Personal Religion and the Life of Devotion, 1924; The Platonic Tradition, 1926; England, in Modern World Series, 1926; Lay Thoughts of a Dean, 1926; The Church in the World, 1927; Assessments and Anticipations, 1929; and some school books.

II

DEAN INGE

I HAVE heard Mr. Lowes Dickinson's lecture, and I agree with parts of it. Not long ago he wrote, if my memory serves me, "For myself, I am no democrat". Now he says he is one. It all depends on what you mean by democracy. In America it means anything you like, but when the word is mentioned you are expected to salute the star-spangled banner. For instance, an American divine exclaimed, "You cannot separate God and democracy. For if we believe in God, we believe in God's purposes, God's ideal, and that is believing in God". The logic seems to halt. Or I may cull this gem from the *New York Medical Journal*, in an article on gout: "Uric acid is tottering upon its throne. Democracy is advancing in medical theory as well as in political practice."

Well, it is bad manners to smile at our friends when they are at their devotions. But, as a matter of fact, democracy is neither an attribute of the Deity nor a method of therapeutics. It is the name of an experiment in government. During the war we said we were fighting to make the world safe for democracy. That was a lump of sugar for the American eagle, and, fortunately for us, he swallowed it. At present, most of the world seems to have made up its mind that democracy is not safe for itself.

We must, however, distinguish between democracy as a form of government, in which all the citizens legislate without representation; democracy as a form of State, which means that the electors, under universal suffrage, have the last word; and democracy as a form of society, which means equal consideration for all. The first is possible only in a small city State, like ancient Athens or a Swiss canton. The second is what we have got. The third is a Christian principle, and as a Christian I believe in it. Number two I do not much believe in, and I fancy very few people believe in it any longer. Mr. Shaw has said that the great political problem is to find a good anthropometric method, and that we have not found it. The silliest of all methods is to break heads; the next silliest is to count them. Practically, universal suffrage means that the worldly goods of the minority are put up to auction at each election, and there is no limit to the absurdity of the promises made by candidates except the fear that they may be called upon to redeem them. It is a ridiculous arrangement; but I frankly admit that I do not know what we could put in its place.

As for Mr. Dickinson's idea of abolishing private gain, I will only make one obvious remark. If you destroy the chief motives which induce people to work hard, namely, the desire to improve their own position, and still more to give their children a good start in the world, a few people will work as well as they do now (I hope I should, but I doubt it), the majority will work badly, and a considerable number will refuse to work at all unless someone stands over them with a whip. The output of commodities would beyond question be enormously reduced; and the country would be very poor. At last in desperation we should adopt the whip, or some equivalent. As Herbert Spencer said, "Socialism would mean slavery, and the slavery would not be mild".

I do not want to give all my space to politics, but one thing I must say before I leave the subject. Those who go about abusing our social system, calling it a hell upon earth and so forth, are doing about the worst disservice to their country that any man could do. For, in spite of all the faults that may justly be found with it, it is a simple fact that there has never been a time in the history of the world when the average citizen, the working man and woman, could command anything like the comforts and amusements and opportunities for education and intellectual pleasure that he and she have now. With all its faults, the civilisation of the twentieth century is the happiest and best for the average man and woman that the world has seen. And then these gentry go about poisoning people's minds, and stirring up discontent everywhere, till the temper of society becomes sour, embittered, despondent, and quarrelsome. Our social arrangements are better for the poorer citizens than they have ever been before, and they are in the way to become better still. America leads the way. By mass production on a large scale, by standardisation, and by improved machinery, rough manual toil is being eliminated, and comforts hitherto undreamed of are being put within the reach of almost all, without lengthening the day's work or diminishing wages. This is the American alternative to Socialism; it *works*, whereas Socialism has always been a dead failure.

There is, of course, a very awkward snag in our way—unemployment, due partly to the dole and partly to over-population. Here I hope Mr. Wells may say something to support me. Our social problems can be solved if our numbers are properly regulated; if they are not, they are hopeless. But my studies in this subject lead me to think that in this country we shall adjust our population to something like the optimum number, though we may go through a bad time first. My calculation is that between 1940 and 1945 the population will become stationary. If it is too large now, and I think it is, the best remedy is State-aided colonisation. The dole is utterly demoralising; its chief effect is to turn the unemployed into the unemployable.

I am more anxious about quality than quantity. The young science of eugenics is going to be extremely important some day; but I am not in favour of very drastic measures to stop the procreation of the unfit, till we know rather more than we do now. But let us all agree that the test of the welfare of a country is the kind of men and women that it produces; and that nature is more important than nurture. There are two great factors in our modern civilisation. One is industrialism and the growth of applied science; the other is the scientific faith.

I am not a man of science, as Mr. Haldane will perhaps remind you; but I read scientific books, and I think I understand the scientific spirit. On the whole, I think the moral influence of the new knowledge has been beneficial. The air that blows round science is like the air of mountain tops, cold and thin, but pure and bracing. I will mention some of the gains which the scientific temper has brought us.

Even in politics and religion, where passion and prejudice are most potent to obscure the intellect and distort the judgment, there is a higher standard of veracity and more respect for evidence. Rhetoric and advocacy are distrusted. The scientific spirit has transformed history, and has imposed rather more conscientiousness even upon controversial literature and public speaking.

Curiosity, which was condemned by monkish morality, is now praised, as it was by the Greeks. To seek for the truth, for the sake of knowing the truth, is one of the noblest objects that a man can live for. Huxley thus states the aims of his own career: "To promote the increase of natural knowledge, and to forward the application of scientific methods of investigation to all the problems of life, in the conviction that there is no alleviation of the sufferings of mankind, except veracity of thought and action, and the resolute facing of the world as it is, when the garment of make-believe is stripped off."

The centre of gravity in morals, as in theology, is changing from authority to rational motive and the conscience of the individual. New moral demands arise from new knowledge and new circumstances, and these new demands are easily stifled by authoritative tradition. Among the big questions which the new morality will have to tackle are our duty to posterity, our duty to the so-called lower animals, and our duty to our habitation, the earth, the beauty of which we are spoiling as fast as we can. We shall have to fight the politician, who remembers only that the unborn have no votes and that since posterity has done nothing for us we need do nothing for posterity; the traditionalist theologian, who tells us that the animals have no souls, and therefore no rights; and the Philistine who cannot see a waterfall without wishing to turn it into a power station.

Again, although science is for the most part

agnostic about the existence of a personal God, it is positive in rejecting much that has been falsely taught and believed about God. God is, at any rate, not a capricious and cruel Oriental sultan, nor a magnified schoolmaster, nor the head of the clerical profession. This purification of the idea of God is a great gain. Such as men themselves are, such will God appear to them to be; and such as God appears to them to be, such will they show themselves in their dealings with their fellow men.

The abandonment of miracle, as a fact of present-day experience, is a clear gain. Though we are still plagued with priestly frauds and bogus cures, ghostly apparitions, and superstitions of every kind, science has laid the axe to the root of the tree, and we may hope that by degrees such beliefs and half-beliefs will either be discredited or placed on a scientific basis.

Lastly, the greatly extended horizon which science has opened for the human race gives us "the rapture of the forward view" which we never enjoyed before. The secular faith of to-day is the belief in rational self-determination, the hope that humanity has its future in its own hands.

So much in praise of science. It does not follow that we must adopt the very poor philosophies which scientific men have constructed. In philosophy they have much more to learn than to teach. The notion that the real is what can be weighed and measured, and that all our higher interests are a kind of luminous haze floating above the real world and unable to affect it at all is very bad philosophy, and theology is quite right to protest against it. It would leave us with no art, no religion, and no science either. The eternal and absolute values are at least as much parts of reality as atoms and electrons.

Nor need we accept the very unscientific superstition of the nineteenth century, the belief in an automatic law of progress. It was held by most of the scientists, but it is quite unscientific. Progress is a rather rare phenomenon in nature and it cannot go on for ever. When we look at the moon, airless, waterless, cold, and dead, we know what the ultimate fate of our planet must be. "But we have a very long lease, almost as good as a freehold?" Yes, we have; but there is no law of progress. When we think of the insect civilisations, which advanced to a more complete socialism than even Mr. Shaw has pictured, we may infer that at least a possible fate for us is to reach a condition of stable equilibrium, when the faculty of thought, which will be no longer useful, will be withdrawn, and instinct will take its place. We shall be in purgatory, but we shall not know it. We shall then have a Socialist and Suffragette millennium, governed by our maiden aunts, who will be armed with stilettos to extinguish the men when they have performed the only function which will be left to them. This may be the end of nature's queerest experiment, the evolution of our noble selves. However, since the human race does not possess the virtues of the little busy bee, and of the ant who is an example to the sluggard, and since all our progress has been the result of our fixed propensity to live beyond our incomes and save ourselves trouble, it is more probable that we shall go on indefinitely as we are, trying new experiments, all of which will be interesting, and some of them successful.

You will expect me to say something about religion in an age of science. Christianity, after breaking the first moulds into which the precious metal, still hot and liquid, was poured, congealed and petrified—you may spell petrified with a capital P if you like—at a rather unhappy period of the history of Europe. The Greeks and Romans knew that their civilisation was on the wane, and they had no hopes for the future.

The Jews had hopes for the future, but what they hoped for was a supernatural deliverance, which would enable them to bruise their oppressors with a rod of iron. The consequence was that neither from Palestine nor from Greece could the early Christians get any evolutionary doctrine of history. The framework of Christianity was catastrophic. There was no significance in history apart from a few tremendous events in the past and future—the creation of the world a few thousand years ago, the fall of man, his redemption in the first century A.D., and the end of the world, which was thought to be very near. We now live in an age when all educated people believe in a very different world order. We are now told that the sun has been in existence about eight billion years, the earth about two thousand million, the human race about one million; and that there is no reason, so far as we know, why there should not be men and women on this planet a million years hence. Further, we believe that the changes in the world are slow and gradual, and in accordance with natural laws. The supreme question for Christians is whether the catastrophic scheme which we have inherited by tradition can be fitted into the evolutionary scheme in which we have come to believe. Personally I think it can, but only at the cost of greater changes than most churchmen are willing to face.

As Protestant Christians, we are bound to the New Testament, and the New Testament only. We are not bound to accept the extreme asceticism which captured the Church—it was not originally a Christian movement—nor the theocratic monarchy which established itself on the ruins of the West Roman empire. I can accept the dictum of Rudolf Eucken, a great German thinker: "We not only can be, but we must be Christians; only, however, if we recognise that Christianity is a progressive historical development still in the making."

There is only one omission in the moral teaching of the New Testament, and I have already indicated what it is. There is no vision of an earthly future either for State or Church, no glimmering that there might be a very long time before the predicted end of the age, no help towards constructing a better social order. When Christ was asked to arbitrate in a case of disputed property, He replied that it was no business of His, and added, "Beware of covetousness".

I have explained this omission; but it is really a good thing. If Christ had laid down laws for the better government of society, He would not only have done no good in the state of society then existing, but his rules would have been a positive hindrance to later societies, whose conditions are totally unlike those of Palestine in the first century. He gives us broad principles, and has a good deal to say about selfish and wasteful consumption; as for problems of distribution, He leaves them for Christian governments to settle in their own way. It is really a great gain.

Christianity, however, is a religion of spiritual redemption, not of social reform. Christ cared very little for the paraphernalia of life. He lived on a higher plane, in the conscious presence of His Father in Heaven. And the religion which He meant to found was a religion of the Spirit, a life of purity and holiness, of faith and love, a Church from which no one is excommunicated except by himself, a brotherhood of men and women who find nothing hateful except hypocrisy, hard-heartedness, and calculating worldliness. Some people have called it a feeble religion, which glorifies weakness and littleness as such. There could not be a greater mistake. It is a heroic religion; but it has its own standard of values; it has broken down all man-made barriers by ignoring them; and, by making the perfection of the divine life the standard of measurement, has made all differences except moral ones sink into insignificance.

This religion is a permanent acquisition of the human race; it is a treasure which we can never surrender. If anyone thinks that the man of the future will want no religion, he must be a very foolish person. The grandest spirit of Western Christianity, St. Augustine, did not even claim that Christianity was new. He wrote, "What is now called the Christian religion was in existence among the men of old time, and has never been lacking since the beginning of the human race, till Christ Himself appeared in the flesh. Since that time the true religion already in existence began to be called the Christian religion". What happened at the Incarnation was that the eternal Christianity appeared for the first time upon the plane of history, and became a power in the world. To suppose that we shall need no religion is nonsense. And as for starting a new religion, as some suggest, you might as well try to build a tree.

Do you object, as evolutionists, to the perfect character being supposed to have appeared nineteen hundred years ago? Then listen to Rodin, the great French sculptor: "In art there is no law of progress. Beyond Pheidias sculpture will never advance." So in the sphere of character we may say, "Beyond Jesus of Nazareth man will never advance".

I base my faith in Christianity mainly on two things. First, the testimony of my heart and conscience (and, may I not say, of the heart and conscience of all right-minded people?), that in the New Testament are to be found "the words of eternal life"; for, as one of the Cambridge Platonists said, "Christianity is a divine life, not a divine science". And my second ground is what is called mysticism or personal religious experience. St. Paul's faith was based on communion with the Spirit-Christ, an experience which was to him absolutely certain. I cannot have his tremendous conviction, for I have not deserved it. Spiritual things are spiritually discerned. But I think I am sure that when I pray in the name of Christ I am not merely talking to myself, or practising moral dumbbell exercises.

So much, then, for my views about religion. I think we shall have to let some traditional dogmas go, but I do not think that will matter. The worst enemies of Christianity are bad Christians; and if I had to find a name for that view of life which is incompatible with Christianity, I should call it *secularity*, that which the New Testament calls "the world", human society as it organises itself without thought of God and the spiritual life.

Some of you may have expected that I should dwell more on the dangers that lie ahead of us. Ten years ago I did think that civilisation was in great danger, and I still think that it was. Now I am inclined to think that we have turned the corner, but we are not out of the wood yet. Another war, or a revolution, might yet plunge us into another dark age.

But I do not want you to confound reasoned hopefulness with optimism. An optimist is a barometer stuck at Set Fair, whatever the weather may be. The man who says, "I am always an optimist", is a very irritating kind of fool. He is the kind of man who would buy from a Jew and sell to a Scot and expect to make a profit.

I repeat that there is no law of progress. Our future is in our own hands, to make or to mar. It will be an uphill fight to the end, and would we have it otherwise? Let no one suppose that evolution will ever exempt us from the struggles. "You forget", said the Devil, with a chuckle, "that I have been evolving, too".

III

H. G. WELLS

FROM WHO'S WHO (1930)

WELLS, HERBERT GEORGE, B.Sc. Lond.; novelist; *b.* Bromley, Kent 21 Sep. 1866; *s.* of late Joseph Wells, professional cricketer; *m.* Amy Catherine Robbins (*d.* 1927); two *s.* *Educ.:* private school, Bromley Kent; Midhurst Grammar School; Royal College of Science. *Publications* Select Conversations with an Uncle, 1895; The Time Machine, 1895 The Stolen Bacillus and other stories, 1895; The Wonderful Visit, 1895 The Island of Doctor Moreau, 1896; The Wheels of Chance, 1896; Th Plattner Story and others, 1897; Certain Personal Matters (essays), 1897 The Invisible Man, 1897; The War of the Worlds, 1898; When the Sleepe Wakes, 1899; a revised edition of this under the title of The Sleeper Awakes 1911; Tales of Space and Time, 1899; Love and Mr. Lewisham, 1900 The First Men in the Moon, 1901; Anticipations, 1901; The Discover of the Future (lecture to the Royal Institution reprinted as pamphlet) 1902; The Sea Lady, 1902; Mankind in the Making, 1903; Twelve Storie and a Dream, 1903; The Food of the Gods, 1904; A Modern Utopia, 1905 Kipps, 1905; In the Days of the Comet, 1906; The Future in America 1906; This Misery of Boots, a tract in favour of Socialism, 1907; Nev Worlds for Old (an account of Socialism), 1908; First and Last Things, confession of faith, 1908; republished, revised, 1917; The War in the Air 1908; Tono Bungay, a novel of contemporary life, 1909; Ann Veronica 1909; The History of Mr. Polly, 1910; The New Machiavelli, 1911; Floo Games for Children, 1911; Marriage, 1912; Little Wars, a floor gam book, 1913; The Passionate Friends, 1913; The Wife of Sir Isaac Harman 1914; An Englishman looks at the World, 1914; The World Set Free, 1914 The War that will end War, 1914, and The Peace of the World, 191 (war pamphlets); Boon (under the pseudonym Reginald Bliss), 1915 Bealby, 1915; The Research Magnificent, 1915; What is Coming? 1916 Mr. Britling Sees it Through, 1916; The Elements of Reconstructio (under the pseudonym D. P.), 1916; War and the Future; God, th Invisible King; The Soul of a Bishop, 1917; In the Fourth Year (Leagu of Nations); Joan and Peter, 1918; The undying Fire, 1919; The Outlin of History, first published in fortnightly parts and then in several boo editions, 1920, is an attempt to reform history-teaching by replacin narrow nationalist history by a general review of the human record Russia in the Shadows, 1920; The Salvaging of Civilization, 1921; Th Secret Places of the Heart; Washington and the Hope of Peace; A Shor History of the World, 1922; Men like Gods, 1923; The Story of a Grea Schoolmaster (F. W. Sanderson); The Dream, 1924; A Year of Pro phesying, 1924; Christina Alberta's Father, 1925; Collected Works (Atlan tic edition), 1925; Mr. Belloc Objects to the Outline of History, 1926 The World of William Clissold, 1926; Democracy under Revision (Sor bonne lecture), 1927; Meanwhile, 1927; The Book of Catherine Wells 1928; The Way the World is Going, 1928; The Open Conspiracy, 1928 Mr. Blettsworthy on Rampole Island, 1928; Common Sense of World Peac (Address in Reichstag), 1929; The Science of Life (with Julian Huxley an G. P. Wells), a companion to the Outline of History, 1929; The Kin who was a King (Film Synopsis published as book), 1929.

III

H. G. WELLS

IT has exercised my mind a lot to find out how much I could tell you of my Point of View in half an hour. Because I suppose that means telling what I think I am, why I exist, what I think I am for, what I think of life, what I think of the world about me, and things like that. These are questions to which I have given innumerable hours, in conversation, in reading and writing, in lonely places, and particularly in that loneliest place of all, the dark stillness of the night. Is it possible to give you something like a quintessence? Anyhow, I am going to try.

In the perfume factories of Grasse in Provence they show you little bottles of concentrated extract. In this little bottle, they tell you, we have condensed the scent of half a million roses, in this, acres and acres of jasmine. In this brief talk to-night I am trying to give you the gist of many thousands of nights and days of thought. I will try to make myself as clear as possible, but you must forgive me if now and then I have to be more concentrated than explicit.

We have already had the Points of View of Mr. Lowes Dickinson, of the Dean of St. Paul's, and Mr. Bernard Shaw in this series of talks.[1] I will not

[1] See note, p. 21.

spend very much of my time discussing what they have said so well.

Mr. Lowes Dickinson talked of democracy. Democracy, I thought, was the name used rather confusingly for two different methods of government, one used in the little city States of the past and the other in the big States of the present, and I have made and written various criticisms of democracy on that assumption. But Mr. Lowes Dickinson said it meant the fullest freedom of speech and discussion and a respectful treatment of all one's fellow human beings, and I have nothing but agreement with that sort of democracy. And the Dean of St. Paul's talked of Christianity. Christianity I have always held to be defined by its creeds, and since I cannot believe in many statements in these creeds —the Resurrection of the Body, for example—I have always refused to call myself a Christian, because that might have been sailing under false colours. But when the Dean of St. Paul's explains that these creeds do not bind him and that Christianity can learn and alter its ideas without limit, almost am I tempted to call myself a Christian and accept his teaching. Mr. Shaw talked of this and that in a manner that was highly provocative. But I have long since trained myself not to be provoked by Mr. Shaw, and I continue to admire beyond measure the beautiful prose he talks and writes and his admirable pronunciation. He said that Russia is the only country which is training its next generation to be better citizens than this one is. Well, I wonder where he found that out!

If I have a general criticism to make of my three able and distinguished predecessors, it is that they have given us *views* rather than a statement of their *point of view*. They told us what they saw, but not where they stood. Now I want to tell you where I stand. I am unable therefore to join on what I have to say to what my predecessors have said. Instead, I propose to begin at quite a fresh point, and a rather more fundamental one.

I can say best what I have to say by talking first about immortality. I will open my matter with a question. Here is a voice talking to you. Here are thoughts being presented to your mind. This kind of mental intercourse which the wireless makes possible is at once extremely detached and extremely intimate. There has never been quite this effect of impersonal nearness before. We are sharing ideas. Our mental lives are in contact. The question I would put is: How far can we consider this mental life we are sharing to be immortal? And more particularly I would ask you a question I have often asked myself. What is this H. G. Wells who is now thinking before you and with you?

Now what do you suppose our little conference amounts to? What is happening now? You are Mr. So-and-so, or Mrs. So-and-so, or Miss So-and-so, and someone called H. G. Wells is talking to you. That is what most people will call self-evident fact. That is what will pass muster as the truth of the matter. But is it altogether true? Let us go into things a little more precisely. I will talk about my side of the talk, which is this H. G. Wells, but what I have to say will apply quite as well to your side also. This H. G. Wells is a person who was born in the year 1866 and who has since gone here and there and done this and that. His voice is here, some thought that may be considered to be his is here, but are you sure that all of him is present here? May I point out to you that so far from all of him being present in this discussion, very much of him is not present anywhere. The greater part of him is no longer in existence. It is dead. It is past and forgotten. He is already for the most part as dead as his grandfather.

Let me explain a little more fully what I mean by this. Consider the childhood of this person. I will tell you of one incident in it. In 1867 he was a small and extremely troublesome infant. He felt things vividly and expressed himself violently. He had one day a great and terrible adventure. It must have seemed like the end of the world to him. He was lying on a sofa and he rolled about upon it and fell off it. He must have been scared by that fall. But also he fell on a glass bottle. It broke. He was cut very dreadfully about the face. This body I have with me to-night still bears a scar over one eye. No doubt he was frightened and hurt, taken up and soothed. The doctor came and sewed him up. What a storm of feeling, what a fuss it must have been! Yes, but what do I know of all that now? Nothing, nothing except what my mother told me of it; nothing else at all. All the fear, all the feeling, all the details of the event have gone out of my conscious existence. All that is quite dead. Now, can I really say that H. G. Wells of one year old is here? You will say, perhaps, "Of course he is". There is the scar. And if that child of twelve months old had not existed, how could this present talker exist?

But wait a moment. That grandfather of mine! He was a gardener and he was rather good at growing roses. One day, towards the end of the reign of King George III, he stood in the sunshine in a garden at Penshurst and budded a rose. I know that for a fact, just as completely as I know for a fact that H. G. Wells fell off a sofa in 1867. And also, be it noted, if my grandfather had not existed the present talker could not exist. My nose and my eyes would not be the shape and colour they are. If the scar is H. G. Wells of 1867, the eye is Joseph Wells of 1828. So, by the same test, if that infant H. G. Wells is alive here, his grandfather is alive here, and so far as one is dead and forgotten, so is the other. There is the same physical continuity; there is the same forgetfulness.

Now this idea that the thinking that is using the voice of H. G. Wells to talk to you to-night is not all of H. G. Wells is a very important idea in my point of view. It is not only that I who am speaking am not in any real sense that baby of 1867, but it is also that I am not a certain ill and angry young man of twenty who lived in 1886. He was struggling in the world under what he thought was an unjustly heavy handicap, and he talked and he wrote. I have photographs of him as he was then; I have stuff that he wrote. And for the life of me I cannot identify my present self with him. I have left him behind almost as completely as I have left my grandfather behind. On the other hand, I have recently been collaborating with one of my sons. We share many ideas and we have very similar mental dispositions. I feel at present much more closely identified with him than with that young H. G. Wells of 1886: or even with the H. G. Wells of 1896, whom I find from a photograph wore side whiskers and a cascade moustache and rode about the countryside on a bicycle.

And now let us turn to another aspect of this curious inquiry. This train of thought which is talking to you now is something very much less than H. G. Wells who is, from my point of view, already very largely dead. But also it is also something very much more than H. G. Wells. You and I are thinking about what is immortal in ourselves. Now H. G.

Wells never started that topic. It came to him. He heard people talking about it and preaching about it. He read about it. People who died in Egypt five thousand years ago and whose names and faces and habits and sins are utterly forgotten were talking about it. Plato, Buddha, Confucius, St. Paul have all had something important to say on the matter. That discussion came into *our* lives as we grew up. We may participate in it, change it a little, before we pass it on. It is like a light passing through a prism which may test it, refract it perhaps, polarise it perhaps, and send it on again *changed*. We are the prism. The thoughts existed before we were born and will go on after we are finished with altogether.

Now here you see is something more, very fundamental, of what I am trying to say to you. Either this will seem the most lucid of realities or the most fantastic of speculations. But first let us have what I am putting to you plain. Here, I say, is this H. G. Wells who is talking, and he is—I have tried to show—so far from being immortal that the greater part of him is already dead and gone for ever. I will not presume to apply the obvious parallel to you. That is *your* affair. But also over and above this H. G. Wells is something, a living growth and a continual refining of ideas, a thought process which is bringing our minds together, expressed by his voice and carried far and wide in radiations from this centre in London. And this thought process has lived already thousands of years ago and may, so far as we know, passing from mind to mind and from age to age, continue its life for ever. We are mortal persons responding to the advance of perhaps immortal ideas. We are not ourselves only; we are also part of human experience and thought.

I hope I have made my meaning clear thus far. You may not agree with me exactly, but I hope you have understood me, so that I can go on to the next light upon my point of view.

A second very fundamental question which man has been debating with himself for many centuries, and which comes to most of us in due time and perplexes us, is the question of what is an individual. It is a question that joins on very closely to these ideas about immortality. How is the individual related to the species? How is the part related to the whole? How is the one related to the many? How is he or she as a whole related to everything in his or her make-up? A great part of the dialogues of Plato, for instance, consists of experiments and explorations about this group of questions.

I agree that to a lot of people this sort of discussion will seem hair-splitting, tedious, and unmeaning. They will fail to see what it is about and what good it is. They feel sure they are individuals, and that is an end to the matter. They will say that they do not want to bother their heads about it. Quite a lot of people seem to live now chiefly to escape having their heads bothered about anything, but most of that kind have probably stopped listening to this quite a while ago, if ever they began. To many, however, these questions are full of meaning, and to some of us they are among the most important questions in the world. They are so to me, and I cannot explain my Point of View at all without discussing them.

I suppose the ordinary and obvious answer to this question of what is an individual would be to say it is a living being detached from the rest of the world. It is born or hatched as a definite distinctive self; it maintains itself for a certain time against the rest of the universe, and at last it dies and comes to at least a physical end. But is that an impregnable statement? If one pries into descriptive biology or into modern psychology, one finds first one curious fact and then another coming up to weaken and undermine this idea of the complete integrity of individuals. They are not so definitely marked off as we are disposed to think.

Go first to the biologist. He will agree that men and cats and dogs are very individual creatures. He will probably say that they are strongly individualised. But when you ask him if that is true of all living things he will at once say "No". He will tell you that most plants seem much more individualised than they are. You can take a plant and break it up into a number of plants. Are they new individuals or are they fractions of the old one? You can take two plants even of different species and graft them together. What is the grafted plant, a new individual or one or both of the old ones? Trees seem to be much more individual than they really are, just as mountains do. It is a disposition of our minds to think of them as individuals. We talk of the Jungfrau or the Wetterhorn as though they were as complete and distinct as pyramids, but really they are only peaks on a general mountain mass. And it is not only plants and all the vegetable kingdom that are wanting in individuality, but the biologist will tell you of innumerable species of lower animals also, of which two sometimes come together and coalesce into one and one will break up into two or many, and again of individuals that branch off others but never separate and so become what are called colonies, a sort of super-individual. If the higher animals could do as the lower animals do, we should have Mr. Lloyd George coalescing with Mr. Snowden into one individual—which I am sure would be a terrible nightmare for the publicists of France—and we should have Mr. Winston Churchill breaking up into dozens and scores of Winston Churchills and writing books, painting pictures, forming governments, commanding and constituting armies and navies, and carrying every aspect of his versatility to the last extreme. I am afraid he would insist upon it. But the biologist assures us that all the higher animals have lost these powers of combining and dividing and spreading themselves out. They are highly individualised, he says; they are unified and drawn together, they are cut off from the rest of the universe into themselves, to a degree no other creatures have attained. These individualities such as we have are an exception and not the rule among living things. They are not the common way of life.

But though we are highly individualised, says the biologist, our kind of creature is not completely individualised. He will tell you of various curious cases when sheep and cats and dogs and babies have been born with two heads to one body or two bodies to one head. When there are two heads, where is the individual then? And he will bring home to you the fact that a great part of our bodily selves is unknown to us. We do not know what is inside of us until we learn about it from talk and lessons and books, and unless trouble is brewing we do not know what goes on inside there nor how it feels. Our particular individuality in fact does not penetrate to our interiors. And if you will let the biologist run on he will tell you that in the blood vessels and substance of our body are millions of little beings, which are extraordinarily like some of the smallest, lowest microscopic animals which lead independent lives, and these go about in our bodies as citizens go about in the streets and houses of a city. These little beings, these corpuscles, kill disease germs, carry food and air about, and do a multitude of services. They have minute individualities of their own. We are made up of millions of such minute creatures, just as cities and nations are made of millions of such beings as we are. There are, you see, different ranks and kinds of individuality. It is not the simple matter so many people assume it to be.

Now when we turn from the modern biologist to the modern psychologist we get still more remarkable revelations about this individuality of ours, which seems at first so simple. He tells us of minds split and divided against themselves. I do not know whether you have read of cases of what is called divided personality. They are fascinatingly strange. They are rare but they occur. There are people who suddenly forget who they are. The individual becomes someone else. That may happen under hypnotism. It may happen in cases of insanity. But it may also happen without either hypnotism or insanity. In the same brain and in the same body it is possible for first one and then another personality to take control. Perhaps you have read a story of R. L. Stevenson's which was suggested by these cases: the story of *Dr. Jekyll and Mr. Hyde*. That puts these phenomena in an extreme fantastic fashion and it ascribes the change-over to a drug. But the change in the actual cases occurs without a drug. Quite a number of us go some little way towards such a change. Which of us indeed has not a better self and a worse self?

I have had to make this appeal to biology and psychology with my eye on the clock, but I think I have at least said enough to show you the support I find in these sciences for my profound doubt whether this H. G. Wells of mine is really the completely independent, separate, distinct being, that it is our habit of mind to consider him. Perhaps my individuality, my personality, seems to be distincter than it is. Perhaps it is—how shall I put it?—a convenient biological illusion.

If I had the time I could produce a great mass of facts to support that, to show how individuality has arisen in the course of evolution and how every individual is, as it were, a sort of experiment made by nature to test this and that group of qualities. In collaboration with Julian Huxley and my son, G. P. Wells, I have been trying to present that mass of facts to the general reader in a work called the *Science of Life*, but our utmost efforts to compress and simplify leave us with a large book, and so I can only allude to it here as being full of light upon this issue, the sort of light there is no time to give you now, and then turn to another aspect of this question of "What am I?" and "What are you?"

Let us look within. How do you feel about your identity with yourself? Well, anyhow, let me tell you how I *feel* about H. G. Wells. I have already tried to show that as a matter of fact a lot of him is already dead stuff and irrelevant stuff, and I have also tried to show that this *thought* that is talking to you is something very much more than H. G. Wells. And when it comes to introspection, then I feel, very, very clearly, that I am something very distinct from this individual H. G. Wells, who eats and sleeps and runs about the world. I feel that I am linked to him as a boat may be moored to a floating buoy. More than that, I have to use his voice, see with his eyes, experience the pain of any physical misfortune that comes to him. He is my window on the world and my mouthpiece. I have to think in his brain, and his store of memories is my only reference library. I doubt if I can think or feel or act as an individual without him. But I do not feel that I am he.

I take a great interest in him. I keep him as clean as I can and am always on the watch to prevent him getting sulky, dull, or lazy—not always with success. He has to be petted and persuaded. I like to be told he is good and remarkable, just as I like to be told my automobile is a good one. But sometimes I wish I could get away from him—heavens, how I wish it at times! He is clumsy in all sorts of ways, and unbeautiful. His instincts and appetites are dreadful. He begins to show considerable signs of wear. The reference library in him might be better arranged and the brain cells quicker at the uptake. But he is all I have to keep me in touch with the world. When he goes, I go. I am silenced for ever.

Now there is nothing original in this sense of detachment from myself. Most people get to something of the sort. When we are young we identify ourselves with ourselves very completely and fiercely. That may be a biological necessity. But as we ripen —or as we age—the separation widens. All through the historical past of our race one can trace this feeling of detachment. They used to call the part that is talking to you now the soul, and the rejected part the body; but that is not quite my Point of View. The H. G. Wells I look down upon is mental just as much as he is physical; he is the whole individualised self-centred personality. When I read St. Paul and find him talking of the Old Adam and the New Adam, he seems to be saying something very much nearer to the truth than that popular distinction of body and spirit. When he cries, "Who can deliver me from the body of this death?" I find him very understandable. How warmly have I echoed that cry! My feeling is just that sense of being *thought*—a part of a great process of thought—which finds itself entangled—as some young creature may be entangled in its egg membranes—in an over-developed, over-intense, over-limited egotism.

Now what I am saying here is not, I believe, an orthodox Christian view. Orthodox Christianity insists that we are ourselves for ever and ever. Mr. Gilbert Chesterton ought to tell you about that. My Point of View is much nearer stoicism. It is indeed stoicism seen in the light of modern biological science. I do not believe in the least that either the body of H. G. Wells or his personality is immortal, but I do believe that the growing process of thought, knowledge, and will of which we are parts, of which I am a part and of which you are a part, may go on growing in range and power for ever. I think that Man is immortal, but not men.

There you have my Point of View, given to you as precisely and clearly as I can. Man, I take it, man in us, is more important than the things in the individual life, and this I believe not as a mere sentimentality but as a rigorously true statement of biological and mental fact. Our individuality is, so to speak, an inborn obsession from which we shall escape as we become more intelligent. And we are under a necessity to escape from it as we become more intelligent, because increasing intelligence brings us more and more clearly face to face with the ultimate frustration of every individual desire in age, enfeeblement, and death. Personality, individuality, is a biological device which has served its end in evolution and will decline. A consciousness of something greater than ourselves, the immortal soul of the race, is taking control of the direction of our lives.

If I had the time and erudition I think I could make an argument to show that this idea of the immortal soul of the race in which our own lives are like passing thoughts, is to be found in what Confucius calls the Higher Person, in what St. Paul calls the New Adam, in the Logos of Stoics, in the modern talk we hear of the Over Man or Super Man. But I cannot pursue these suggestions now.

But if I may say a word or so about the views one gets from this Point of View, I would insist first that the subordination of self to a higher order of being does not mean the suppression of all or any of one's distinctive gifts. We have to use ourselves to the utmost. We have to learn and make to the full measure of our possibilities. It is a sin to bury the talent, the individual gift which we possess for the good of the master being, Man. Nor must you imagine that the subordination of self to the immortal being of the race means a subordination of one's narrow self to the equally narrow selves of other people. It is for them also to give themselves to that life and all that increases knowledge and power. I do not believe in the surrender of one jot or one tittle of one's intelligence and will to the greatest happiness of the greatest number, or to the will of the majority, or any such nonsense: I am not that sort of democrat. This world and its future is not for feeble folk any more than it is for selfish folk. It is not for the multitude but for the best. The best of to-day will be the commonplace of to-morrow.

If I am something of a social leveller it is not because I want to give silly people a good time, but because I want to make opportunity universal, and not miss out one single being who is worth while. If I want economic change it is because the present system protects and fosters a vast swarm of wasteful spenders, no better in their quality and much worse in their lazy pretentious traditions, than the general run of mankind. If I am opposed to nationalism and war, it is because these things do not merely represent an immense waste of energy, but because they sustain a cant of blind discipline and loyalty and a paraphernalia of flags, uniforms, and parades that shelter a host of particularly mischievous unintelligent bullies and wasters, because they place our lives at the mercy of trained blockheads. Militarism and warfare are childish things, if they are not more horrible than anything childish can be. They must become things of the past. They must die. Naturally my idea of politics is an open conspiracy to hurry these tiresome, wasteful, evil things, nationality and war, out of existence, to end this Empire and that Empire, and set up the one Empire of Man. And it is natural that I should exalt science. In the scientific world I find just that disinterested devotion to great ends that I hope will spread at last through the entire range of human activity. I find just that co-operation of men of every race and colour to increase Man's knowledge. We can all be citizems of the free state of science. But our political, our economic, our social lives have still to become illuminated and directed by the scientific spirit, are still sick and feeble with congenital traditionalism.

My time draws to an end. I was asked to give my Point of View, and I have given it. I hope to have interested you and I hope I have not offended you. This is how I try to live, and this is how I have got to a certain mastery over the greed, the fears, the passions and vanities that troubled my earlier days, and rid myself altogether of the fear of death.

It is good to be a part of life. Just as a sundial only counts the sunny hours, so does life know only that it is living. Many experiences there are in life, but one there is that we shall never have. We shall never know that we are dead. My Point of View, I can assure you, is not an unhappy point of view. I have found it a good working point of view. I wish you—you other fragments of Man—could tell me what you think of it. I wish we could turn this apparatus about now and I could listen in to you.

IV

J. B. S. HALDANE

FROM WHO'S WHO (1930)

Haldane, John Burdon Sanderson, Sir William Dunn Reader in Biochemistry, Cambridge University, since 1922; Head of Genetical Department, John Innes Horticultural Institution, since 1927; Fullerian Professor of Physiology at the Royal Institution, 1930; *b.* 5 Nov. 1892; *s.* of J. S. Haldane; *m.* 1926, Charlotte, *d.* of Joseph Franken. *Educ.:* Oxford Preparatory School; Eton; New College, Oxford (M.A.). Served in Black Watch (France and Iraq), 1914–19 (twice wounded); Captain, 1915; Fellow of New College, 1919–22; Corresponding Member Sociètè de Biologie, 1928. *Publications:* numerous scientific papers on human chemical physiology, genetics, natural selection, and other subjects, mainly in Journal of Physiology, Journal of Genetics, and Proceedings of Cambridge Philosophical Society; Daedalus, 1924; Callinicus, 1925; Possible Worlds, 1927; Animal Biology (with J. S. Huxley), 1927; Science and Ethics, 1928

IV

J. B. S. HALDANE

I DIFFER in several respects from the other speakers in this discussion. To begin with, I am twenty-seven years younger than the youngest of them. So that I am the only representative of the generation whose finest members died in the Great War. And my intellectual background is very different. As a child I was not brought up in the tenets of any religion, but in a household where science and philosophy took the place of faith. As a boy I had very free access to contemporary thought, so that I do not to-day find Einstein unintelligible, or Freud shocking. As a youth I fought through the war, and learned to appreciate sides of human character with which the ordinary intellectual is not brought into contact. As a man I am a biologist, and see the world from an angle which gives me an unaccustomed perspective, but not, I think, a wholly misleading one.

In describing the world as I see it in so short a time, I cannot avoid being dogmatic. I do not doubt that some of the statements which I am going to make are false. A survey of the beliefs which intelligent men in the past have held as certainties makes that sufficiently clear. One cannot order one's life without a set of beliefs of some kind. But the intellectually honest man must recognise the utterly provisional nature of his beliefs. So when I make an apparently definite statement, I must ask you to put before it some such words as "It seems to me very probable that . . . ". I will now try to state my Point of View.

Man lives in two worlds, the visible world which changes with time, and an invisible world whose constituents do not change. But both worlds can only be described as they appear to us, that is, from a human and imperfect standpoint. Among the components of the invisible world are the realities corresponding to mathematical statements like $16 + 9 = 25$. This is a statement of a fact as real as the Albert Memorial, which any sane person must recognise when it is pointed out to him. But unlike the Albert Memorial, it was a reality 10,000 years ago, and will be 10,000 years hence. There are also invisible realities corresponding to scientific laws, and I think also to some of our general notions of what is beautiful and good. These latter realities are harder to apprehend because we approach them through a mist of emotion. We know very little about what may be called the geography of the invisible world. The religions, if I may continue the metaphor, have covered the vacant spaces of its map with imaginary monsters, the philosophies have ruled them with equally imaginary parallels of latitude. But both have affirmed, in opposition to the so-called practical man, that the meaning of the visible world is to be found in the invisible. That has been the secret of their success. They have failed when they tried either to describe the details of the visible world or to dictate the details of conduct in it. The churches are half empty to-day because their creeds are full of obsolete science, and their ethical codes are suited to a social organisation far simpler than that of to-day. But they still command the allegiance of a number of intelligent people because, amid a world of transitory interests, they support in some measure the claims of the ideal. I am not a member of any religious body, because I find those claims upheld elsewhere. If I thought that the aims of science and art were merely material I should belong to some church. But I believe that the scientist is trying to express absolute truth and the artist absolute beauty, so that I find in science and art, and in an attempt to lead a good life, all the religion that I want.

I have not very much use for people who are not in touch with the invisible world. At best they are good animals, and too often not even that. The men and women who have done best, both for themselves and their fellows, are those who have brought these two worlds into relation. For example, you can hear me to-night because James Clerk Maxwell fifty-six years ago embodied an extremely important set of properties of electro-magnetic waves in a set of equations. Those equations represent an eternal truth—something in the invisible world. Their discovery rendered wireless communication possible. If you do not make any contact with this timeless world (in other words, have no inner life) you have at best a very precarious hold on happiness. Given that contact, you enjoy a very considerable security from the results of misfortune in the visible world, and a complete immunity from boredom. Dean Inge, H. G. Wells, and I, agree to a considerable extent about the nature of the invisible world, because we are all, in some degree, disciples of Plato.

One does not come naturally to the realisation of eternal truths and values. One is brought there by education in the widest sense. It is one of my principal functions to teach certain scientific truths to students at Cambridge University. Many of them are both able and eager to learn. But others are neither able nor eager. Under our present economic system they are enabled to come to Cambridge because their parents are wealthy. By so doing they keep out others who are better qualified intellectually to learn, and more willing to do so. As a teacher I cannot support a social system which is responsible for this injustice. We have got rid of physical starvation. We still have intellectual, æsthetic, and spiritual starvation, which to my mind are greater evils than any mere economic inequality.

Until our educational system is so altered as to give a fair deal to every boy and girl who desires a first-rate education and is capable of benefiting by it, my political views are likely to remain, as they are now, on the left.

There is a worse evil than intellectual starvation, and that is the deliberate suppression of free thought and free speech. I rejoice to live in a country where this evil, though it exists, is less serious than in most other countries. But I believe that even in England freedom of publication is unduly restricted in the names of decency, morality, and so on. There is much more liberty in this respect across the Channel, and no one who has worked beside the French in peace and fought beside them in war can accuse them of degeneracy. So many new ideas are at first strange and horrible though ultimately valuable that a very heavy responsibility rests upon those who would prevent their dissemination.

Moreover, the censorship to which I refer is applied in a very partial way. A book glorifying war may be quite as anti-social, and to my mind quite as obscene, as one glorifying illicit love, but it is never suppressed, and seldom publicly denounced.

I now turn from the world of ideas to the visible world. I am a biologist, that is to say, I study the nature of living creatures, and I naturally look at things from a biological point of view. I feel at home in the world because I know that the other animals, and the plants, too, are my blood relations. Even the inert matter has mostly been alive in the past. When I look at a limestone mountain I realise that, grim and lifeless as it appears, it was made by countless billions of my microscopic fellow creatures. What is more surprising, I think that I can even have some very dim inkling of what it feels like to be limestone. We know material objects in general from the outside. We know our own bodies from the inside. Just as everyone knows what it feels like to be hot, so I know from my own personal experience what it feels like to consist of an abnormally large or small amount of calcium carbonate, of which the limestone mountain is built. In this concrete and detailed way I feel my relationship to the world around me.

I am a part of nature, and, like other natural objects, from a lightning flash to a mountain range, I shall last out my time and then finish. This prospect does not worry me, because some of my work will not die when I do so.

As a biologist I am interested in my body. Most people are only interested in anything below their skins when they are ill. I like to study the performance of mine as my friends do that of their motor-cycles or receiving sets. It amuses me to know what my heart does when I run upstairs, or how quickly my finger nails grow. To a biologist even a toothache can be interesting. Naturally I regard health as extremely important, far more so than wealth, and I shall regard my life as well spent if I can do a little, by research and education, to make my fellow creatures healthier. There is still an immense amount to be learnt about health, but if what is at present known to a few were part of the general knowledge, the average expectation of life in this country could probably be increased by about ten years. Two difficulties lie in the way: ignorance and the dissemination of falsehoods. To take a simple example of the latter. Enormous sums are spent in disseminating lies about health in order to advertise medicines and "health foods" which are generally useless and often dangerous. A widely advertised vitamin preparation contains, besides vitamins, a substance definitely poisonous to children. Under the law of the land I might have to pay thousands of pounds in damages if I mentioned the preparation in question, even if my statement could be proved to be true. On the other hand, I am at liberty to say publicly that diphtheria antitoxin is useless, which is a plain lie.

Now for an example of the prevailing ignorance. When a father advises his son on a choice of occupation, he is generally guided mainly by economic, and partly by ethical considerations. He wants his son to avoid bad wages and bad company. He does not think about bad health, though he may be impressed by the risk of violent death. Yet the health of different occupations differs to an extraordinary extent, and the average man knows very little about the risks of even his own job, let alone his neighbour's. Otherwise no sane man would take up such an occupation as that of metal grinder or barman, with a mortality double that of the average man, when he might become a carpenter or a railwayman, and thus enjoy an expectation of life above the average. Our rulers are equally ignorant of these matters. Protective duties and subsidies are granted quite impartially to healthy occupations like agriculture, and unhealthy ones such as the cutlery trade. When this policy is opposed it is opposed on economic grounds, and never because, by encouraging an unhealthy trade, you are condemning some of your fellow countrymen to death. All parties agree in putting economic considerations before biological; wealth before health. I could give you plenty more examples of this ignorance if time permitted.

Even a healthy man or woman is incomplete. For a large number of men the main interest in life, the main object of their desires, the main source of their satisfaction, is Woman. For me the fascination of woman is only second to that of science. In most cases man's interest in woman culminates in marriage. Provided it does not then cease, the marriage is generally a success. Successful marriage requires a certain effort by both husband and wife.

But, speaking as a happily married man, I can assure you that no other effort is so amply rewarded.

COMPARATIVE MORTALITY OF SOME OCCUPATIONS (ENGLAND AND WALES, 1921–1923)

STANDARD MORTALITY FOR CIVILIAN MALES BETWEEN AGES 20 AND 65: (*basis* 100)

Anglican Clergymen	56
Insurance Officials	60
Farmers	67
Agricultural Workers	68
Civil Servants (all Grades)	80
Railwaymen	83
Electrical Workers	85
Motor Vehicle Drivers	86
Carpenters and Woodworkers	88
Printers	95
Metal Workers	96
Salesmen and Shop Assistants	97
Builders	99
Clerks (not Government)	99
Coal Miners	101
Doctors	102
Textile Workers	105
Makers of Clothing (including Boots)	108
Workers in Amusement Trades	121
Brewers and other Makers of Drinks	126
Glass Workers	128
Horse Drivers	138
Potters	148
Dockers	150
Innkeepers	162
Seamen	177
Bookmakers	193
Barmen	196
Tin and Copper Miners	325
Cutlery Grinders	330

Marriage has a biological basis, and would be far more often a success if its biology were generally understood and the knowledge acted on. But you can only study the physiology of marriage against a background of general human physiology. If you do so the facts fit into their proper places. If not, you get a distorted and unhealthy view of them.

The psychological, even the intellectual, benefits of marriage, seem to me to be enormous. If a man has lived for some years in the closest intimacy with a woman, he learns to look at life from her point of view as well as his own. A man who cannot do this is like a man blind in one eye. He does not appreciate the solidity and depth of the world before him. The ideas I am putting before you here are largely my wife's, or at any rate, family ideas, rather than my own private productions. The unmarried woman is perhaps even worse off than the unmarried man; and few women seem to me to be psychologically complete till they have become mothers. During the Middle Ages Europe was far too much influenced by celibate men. To-day much too big a part in public life is played by the celibate woman, and too little by mothers. I find few ideas more genuinely disgusting than that held by many education authorities that a woman ceases to be suitable as a teacher when she becomes a mother. Because I have so high an opinion of marriage at its best, I think that it should be possible to end it if it fails for any of a number of reasons, instead of, as now, for one only. This is called "undermining the sanctity of marriage".

Marriage generally brings children. Everyone will agree that it would be an evil if the birth-rate of this country were halved, in which case the population would rapidly fall; or doubled, in which case it would increase too quickly. But they will disagree whether too many or too few children are born at present. I do not know myself, though I am clear that too many children are born in the slums, too few in the well-to-do suburbs. But we shall not arrive at a sensible solution of the population problem till we realise that it is a question of numbers, like the design of a motor-car or the framing of a budget, and cannot be settled by an appeal to abstract principles alone.

Our present educational system is unjust to children because the majority of them do not get a fair chance, and practically none are taught the truths of science from a human point of view. Science teaching should begin, not with a mythical body in rest or uniform motion, but with the human body. Mine did so begin at the age of three.

Between different men and women there are immense inborn differences which no amount of education can overcome. I do not believe that any training could have made Ramsay MacDonald into Jack Hobbs, or *vice versa*. The ideal society would enable every man and woman to make the best of their inborn possibilities. Hence it must have two characteristics. First, liberty, which would allow people to develop along their individual lines, and not attempt to force all into one mould, however admirable. Second, equality of opportunity, which would mean that, as far as is humanly possible, every man and woman would be able to obtain the position in society for which they were best suited by nature. The waste of human beings under our present system is a far worse evil than any merely economic waste. I believe in democracy because equality of opportunity is impossible where inherited rank or wealth is important, but for no other reason. I do not know what would be the ideal form of government in a community where that equality had been achieved. Democracy appeals to me, not as an end in itself, but as the most hopeful route, at least for England, to a classless society. In a classless society far-reaching eugenic measures could be enforced by the State with little injustice. To-day this would not be possible. We do not know, in most cases, how far social failure and success are due to heredity, and how far to environment. And environment is the easier of the two to improve.

I am a citizen of the British Empire, which includes the great Dominions. My high-brow friends complain that the Dominions have produced little great art or literature. I answer that at least they have done something unique. Before the war the average expectation of life of a baby born in New Zealand was sixty years, in Australia fifty-seven years, in Denmark, the next healthiest country, fifty-six years. England also ran. Since then other countries have caught up to a large extent, but New Zealand and Australia still seem to be leading. I am proud to belong to a Commonwealth which has won the first and second places in the great race against death.

I am also an European, and proud of it. Europe is sick to-day, but it is at least making some attempt to cure that sickness by a federal union of its states. And it still leads the world in science, literature, art, and music. In methods of production the United States are ahead of us, and many Europeans think that we should copy them. Dean Inge believes that the working class in the United States is better off than our own. His opinion is shared in unexpected quarters. When my wife and I were in Moscow last year at a great scientific congress we only saw two propaganda films. One was against alcohol; the other showed the manufacture of Ford cars as an argument for American industrial methods. I take a different view for the following reasons. Though they are still reducing their infantile mortality, since 1921 the death-rate of Americans at every age from thirty upwards has been increasing steadily. Whether as the result of hustle, prohibition, or the spread of medical cults, such as "Christian Science" and osteopathy, which reject the results of science, America is at present heading for death, and not life. Europe has much to learn from America, a little even from Asia, but I do not think that we should imitate either of these continents.

Some of you probably think I have laid too much emphasis on death rates; I have talked about them for two reasons. Firstly, they are the only means we have of comparing the health of two trades or two nations; and I think that there is a very close connection between health and happiness. Secondly, otherwise well-informed people are ignorant of the facts concerning them.

For example Mr. Bernard Shaw, in a recent book, stated that while the lungs of Sheffield cutlery grinders used to be unhealthy, they were now as healthy as those of other people, thanks to Government Inspectors. Unfortunately the latest available statistics show that the death-rates of cutlery grinders, both from consumption and bronchitis, are between seven and eight times those of the general population. A similar remoteness from facts permeates Mr. Shaw's other opinions on medicine and biological science, some of which were given in his broadcast Point of View.[1]

I am an Englishman, and, what is more remarkable, though of Scottish origin, I believe in England. At the present moment our country counts for less

[1] See note, p. 21.

in international politics than during last century. Nevertheless some of our ideas and practices are at present conquering the world. In Moscow, which has rejected the great British invention of Parliament, there was a word which I constantly noticed on posters. It was not "soviet", nor "red", nor yet "revolution", but "phutbol". The same is happening all over the world. Spanish bull-fighters are becoming centre-forwards. German students are taking to football instead of slashing one another's faces. And with British sport goes the ethical code called Sportsmanship, which future historians may perhaps consider a British invention as important as Parliament and Railways. I hope to see British sport conquer most of the world. But I am no narrow patriot, and would welcome a French invasion of the British kitchen.

England is only likely to regain her former preeminence if we can be ten years ahead of the rest of the world in industry, as we were a century ago. We should, of course, reorganise our industries, but other countries have already done so. We shall not regain our place by doing that. We have probably no great undeveloped mineral resources. But we have undeveloped human resources, especially among the children of the skilled artisan class. Our best hope for the future lies in giving them a chance to become Watts and Stephensons.

Finally, I am a human being, a citizen of the

world which applied science is daily unifying. My own profession of scientific research knows no frontiers and no colour bars. Japanese, Indians, and Chinese, as well as Europeans and Americans, are, or have been, among my colleagues. I am naturally in favour of any measures tending to unify humanity and prevent war. But my views as to the best methods of achieving these aims are not informed by sufficient knowledge to be worth stating. For the same reason I am saying nothing about economics.

I am glad that I live to-day and not at any time in the past. In the 4,000 years before about A.D. 1800 civilisation had spread over a gradually widening area, but its quality had not greatly improved. A century ago in England children were hanged for theft, and a married woman could own no property. Neither of these evils existed in Ur of the Chaldees 4,200 years earlier. In the nineteenth century we doubled our average expectation of life, quadrupled our average real wage, and vastly improved our education and morals. This was made possible, in the main, by the application of science. To-day the whole form of civilisation is changing. We are trying unheard-of experiments. The great experiment of Socialism is being tried in Russia and will doubtless be tried elsewhere. We meet with huge and unexpected accidents like the Great War. We shall go on having such accidents so long as our rulers are not merely ignorant of science, but think on prescientific lines. (You will remember how the Kaiser talked of the war in terms of "shining armour", and Mr. Asquith of "unsheathed swords".) We have got to learn to think scientifically, not only about inanimate things, but about ourselves and one another. It is possible to do this. A single mind can acquire a fair knowledge of the whole field of science, and find plenty of time to spare for ordinary human affairs. Not many people take the trouble to do so. But without a knowledge of science one cannot understand current events. That is why modern literature and art are mostly so unreal.

We live in a dangerous age, but an extraordinarily interesting one. History is being made on a vaster and quicker scale than ever before. For humanity as a whole I am hopeful. For England I am only moderately hopeful, though I believe that if we are willing to adapt ourselves to the new conditions of life, we may yet be as great a nation as ever. But even if I am blown to pieces in the destruction of London during the next war, or starved to death during the next British revolution, I hope that I shall find time to think as I die, "I am glad that I lived when and where I did. It was a good show".

V

SIR OLIVER LODGE

FROM WHO'S WHO (1930)

LODGE, SIR OLIVER (JOSEPH), Kt., *cr.* 1902; F.R.S., D.Sc. London; Hon. D.Sc. Oxford, Cambridge, Manchester, Liverpool, Sheffield, Leeds, Adelaide, and Toronto; LL.D. St. Andrews, Glasgow, Aberdeen, and Edinburgh; Hon. M.A. Birmingham, Hon. M.Inst.E.E.; Principal of the University of Birmingham, 1900–19; *b.* Penkhull, Staffordshire, 12 June 1851; *e. s.* of Oliver Lodge, Wolstanton; *m.* 1877, Mary F. A. (*d.* 1929), *d.* of Alexander Marshall; six *s.* six *d.* *Educ.:* Newport (Salop) Grammar School; University Coll., London. Professor of Physics, University Coll., Liverpool, 1881–1900; Rumford Medallist of the Royal Society, 1898; Romanes Lecturer at Oxford, 1903; President of the Mathematical and Physcial section British Association, 1891; President of the Physical Society of London, 1899–1900; President of the Society for Psychical Research, 1901–4; President of the British Association, 1913–14; Albert Medallist of the Royal Society of Arts as the pioneer in wireless telegraphy, 1919. *Publications:* Elementary Mechanics; Modern Views of Electricity; Pioneers of Science; Signalling without Wires; Lightning Conductors and Lightning Guards; School Teaching and School Reform; Mathematics for Parents and Teachers; Life and Matter; Electrons; Modern Views of Matter; The Substance of Faith; Man and the Universe; The Ether of Space; The Survival of Man; Parent and Child; Reason and Belief; Modern Problems; The War and After; Raymond, or Life and Death; Christopher, a Study in Human Personality; Making of Man, 1924; Atoms and Rays, 1924; Ether and Reality, 1925; Relativity, 1925; Electrical Precipitation, 1925; Talks about Wireless, 1925; Evolution and Creation, 1927; Modern Scientific Ideas, 1927; Science and Human Progress, 1927; The Natural History of a Savant (translated from Charles Richet), 1927; Why I Believe in Personal Immortality, 1928; Energy, 1928; The Phantom Walls 1929.

V

SIR OLIVER LODGE

MR. LOWES DICKINSON started the series as a symposium; that has come to imply not a drinking together but a talking together—a discussion where each contributor has heard his predecessors, and can refer to them if he likes, rather than strike out a special line of his own. I shall try to conform to this implication to some extent; but for the most part I hope to take a line of my own, though its direction will be deflected by the widely heard contributions of the last three speakers.

My friend George Bernard Shaw is a man for whom I have high admiration, for he has immense ability and as a dramatist ranks among the supremely skilled. Deep beneath the surface he is, I try to think, intensely serious, but on the surface he is often perverse, and amusing, when not irritating. His talk in this series has been mainly about politics or sociology; and as usual, when he lets himself go, he has decorated his utterance with wit, has distorted it by exaggeration, and, I fear, has disfigured it by occasional exuberances of youthful indiscretion, such as the sentences about equality of income, and Russian schools. He seems to approve of taxation as a means of benefiting the poor, but he well knows that precious little of it goes to the poor. It mainly goes to pay the interest on financial loans, and to provide for the fighting services, past, present, and future. There is an element of absurdity in most things human, when regarded with a humorous eye, so he chaffs about the appointment of Cabinet Ministers and their domination by the Civil Service; but if our democratic system of government were as hopelessly absurd as his representation of it the country would hardly have been worth fighting for. Surely we can admit that our statesmen are trying to do their best under difficult circumstances. Destructive criticism is easy, but not helpful. It is no light task to rectify shortcomings or errors of the past, to set Society on a more wholesome basis, to promote friendly co-operation among all classes, and to hand on to posterity our noble inheritance. We need all the help we can get in our effort at democratic self-government.

One of Mr. Shaw's main theses is that self-government is impossible, that people are incompetent to govern themselves.[1] He takes Abraham Lincoln's Gettysburg speech, and, while admitting government *of* the people *for* the people, denies that government can be *by* the people. Whereas it seems to me that all stable, durable, and satisfactory government is conducted at the will and with the consent of the governed. It matters little whether rules are made and applied by a Parliament, a Committee, or a

[1] See note, p. 21.

Dictator; if the rules are approved and obeyed by common consent, then that is a good form of government. It is rather like the currency: any form of currency will do so long as people agree to accept it. We are mainly governed by mutual consent. So it is that Summer Time has begun and ended now without the slightest trouble or dislocation. A typical case might be called government by "label" or printed instruction, of which we have an example in the tube railway stations. Instead of a policeman at each corner there is a label, and the crowds willingly abide by the instructions, which are for the convenience of all.

Biologists have now discovered what a great part has been played in the evolution of higher species by symbiosis, as they call it, or the art of living together: far more has been done by mutual aid and unconscious service in Nature than by any fierce struggle for subsistence. The fact is that subsistence itself largely depends on mutual benefit and service. This is the dispensation that has made progress possible; this it is that has developed higher forms by gradual evolution from lower forms. Only by mutual service comes an era of peace and prosperity to animal and plant; only thus can they settle down under a beneficent régime.

Government is the art of living together. It can only be carried on by common consent. The conductor of an orchestra has a stick, but it is a symbol of willing co-operation, not of compulsion. The more we are together the happier we shall be. The art of living together is the essence of civilisation. It is for that reason that we drive on a specified side of the road, and to some extent walk on a specified side of the pavement. For the most part a London crowd, or, I suppose, a crowd in any of our congested areas, is composed of well-meaning pleasant people, who behave in a friendly and considerate manner. They are governing themselves in accordance with the instincts of the community to which they belong. Occasionally there are a few wrong-headed idiots who think it funny to take advantage of the absence of coercion to behave in an inconsiderate and obstructive manner, to interfere with people's enjoyment in public entertainments, and to plead the high spirits of youth. Selfish inconsiderateness is no prerogative of youth; and it is to be hoped that Lord Byng will come heavily down upon all such hooligans.

In addition to wrong-headed self-satisfied idiots who think that people will be amused by their antics, there are, strange to say, a few exceptional vicious individuals who cause trouble maliciously. The lifts in the tube-stations, for instance, are sadly defaced by the apparently necessary notice, "Beware of pickpockets". A professional pickpocket is a thoughtless scoundrel who has not enough imagination to realise the abominable inconvenience he may be causing, and the hardships to people no better off than himself. At one time there were misguided enthusiasts who attacked things which are normally left to the honour of the public—a momentary eruption of insanity which fortunately has subsided. Most of our social arrangements are conducted on the principle that everybody is willing to try to do his best, to protect public property, and give way to and help the weak and defenceless. The prevalent state of mind is a good one, and does not need "Dora". Even now what are the police mainly doing? They are mainly engaged in the civil occupation of regulating the traffic: and that is conspicuously done by mutual consent. It is a good example of the art of living together and of government by consent of the governed. Things can be left to the honour of the public on a large scale, as to a great extent they are in the credit system of the City. In business circles there are sharp practices that are not done, save by an outsider. There are conventions that are universally respected; and on that principle commerce goes on. But I venture to think we are still too much governed in the interests of, or rather, as a safeguard against, the professional criminal. Too many precautions are taken for what may never occur; too much may be expended on mere insurance. Crime is not natural to humanity; it is a disease of the body politic. It would be more economical to attack the root of the evil, and eradicate it, than to be constantly taking precautions against the risk of its evil fruits.

More especially is this true in international affairs. The nations now are well disposed and friendly to each other, perhaps as never before in the history of the world. They have at last learned that attack is ruinous—mutually ruinous; they perceive that they must cultivate the art of living together, that they are an assemblage of mutual interests, and that the destruction of one would mean injury to all. We are emerging from troublous times, but we are emerging with a great hope. We have learned that modern war among civilised nations is an anachronism, that mutual destruction and wholesale damage is lunacy, that actual warfare is no longer a pageant or heroic display, but is a damned dirty and disgusting business. Discipline and duty and useful service can be instilled into youth by other means; witness Baden-Powell's admirable achievement of the Boy Scouts; it does not need war to stimulate those qualities. Let us recognise the merit latent in humanity, develop and apply it in the arts of peace. There is scope for heroism in lifeboat service, as well as in the Navy; in mines as well as in trenches; there are plenty of dangerous occupations, and need for disciplined service and self-sacrifice in all manner of employments.

Indeed, I hope that the art of living together will go further than refraining from attack and destructive violence. I hope that we shall freely exchange each other's goods, and trade without the restrictions of customs-houses. If we want temporarily to protect some nascent or harassed industry, let us do it internally, by bounty, in our own country; but externally let us freely receive from all peoples such commodities as they can properly and legitimately supply, and be thankful.

And let us travel more freely, and get to know each other better. Why do we trouble about passports? Is it to assist the police? But how few criminals there are, compared with the multitude of people inconvenienced by this procedure! Most of us are well-meaning people who travel abroad for business or for health or for enjoyment. We need not be herded together in front of an official, and put through a useless sort of ceremony, for the sake of a few rapscallions, who, after all, would most likely have their papers in good order. Let us be governed more clearly in the interests of the majority, the great majority of the well-meaning and the friendly.

We are beginning to ask, What are our expensive armaments for? Are they for seizing other people's goods, and for unprovoked attack? Not at all, we are not professional criminals. Every nation is learning, or has learnt, that in co-operation, and in innocent friendly intercourse, our interest lies, that violence and mutual destruction are insanity and suicide. The limitation of armaments would set free immense resources for peaceful development. Weapons are only wanted now as a police precaution, an insurance against a wrong-headed or criminal outburst in some essentially uncivilised part of the earth, some nation or community that, however otherwise developed, has not yet appreciated the art of living together, and has to be coerced and suppressed. Insurance precautions of that kind are still necessary; the weapon of the boycott may have to be used; but we must not expend all our resources on insurance.

It is difficult for us to judge a criminal when caught; it is too late then. There must be some reason for his criminality, something that has to be remedied, some disease to be cured within the body politic. We spend too much upon armaments, and too little upon investigation and reform. There is so much to be done, so much that is crying out to be done; and every nation pleads poverty, at least in the old world. But there need be no poverty if we learned the art of living together, and practised it in a spirit of mutual good will.

Above all, the old abominable motto that if we want peace we must prepare for war should be relegated to disgrace. That is not the way to secure peace. Think of private life under such conditions. Suppose we were on guard against our neighbour, always expecting him to go out of his way to attack us, and preparing for retaliation if he did. Private life would be impossible under such conditions. Neighbours get on together in friendly amity. The world is too small and too closely knit together, now, for international enmity and suspicion: the nations are at length beginning to realise that they must be neighbourly. Never before have all the nations of the world met together really and truly in the interests of peace: we are living in most hopeful times. I look for a time when the whole human family will govern itself by mutual agreement, the details settled perhaps at some international parliament at Geneva, with power to coerce the recalcitrant and the criminal:

> There the commonsense of most shall hold a fretful realm
> in awe,
> And the kindly earth shall slumber, lapt in universal law.

Aye, it will come. I see the dawn of it already. Humanity has a long era before it; we are not yet civilised. Every speaker in this series has had to find fault with something. There is much that is wrong, but humanity is in its infancy, and in that there is hope. Evolution has not stopped. We have risen thus far in the scale of existence: the average is still rather low; a few peaks of the race show what will be possible in the future; and the attainment of that future is the work of the present and of the coming generations, work to which all men and women of good will are called. It is a work of which only millenniums lie behind us, while myriads of ages lie ahead. It is no time to fold one's hands and say that things are too difficult or impossible. We must wake, and be up and doing; and the first and most pressing demand of our time is the securing and consolidating of international friendliness and peace.

But I must not limit myself to the progress of the race. From my Point of View (which I have already set forth in a considerable number of books, including one just published), the individual is even more important; and in the space that remains I must refer to the interesting utterances of Mr. H. G. Wells and Mr. J. B. S. Haldane.

Mr. Wells's contribution I have not only heard but read: he gave us a biological point of view. I always want to agree as far as I can with any sincere utterance, and his I felt was both able and sincere. It was as clear and as near the truth as can be got by attending to matter alone. He realised that the higher grades of being had attained individuality; that H. G. Wells, for instance, was an individual, contributing his quota as part of a much larger whole in which his contribution would be merged, that the race was the permanent and immortal thing, the individual only temporary, changing indeed during a single lifetime. He likened us to the white corpuscles in the human blood, each with a kind of identity of its own, and going about its business subserving the good of the whole. Or he might have likened us to the bees in a hive, where the whole community is the important thing, to which the individual bee is sacrificed.

In all this there is a great deal of truth. The individual as we know him is a part of a larger self, a self beneath the threshold of consciousness, of which the portion displayed by the organism is but a fraction. I have previously used the iceberg analogy for the part displayed above the surface and the larger submerged portion beneath. But even apart from that, the thought, the idea, the personality, is greater and more permanent than the material encasement, which, as he says, keeps on changing, has its day, and ceases to be. But the personality does not cease; it is continuous, and is not dependent on the material instrument of manifestation, which it uses for certain purposes, just as we may use the transmission apparatus of the B.B.C. I can go nearly all the way with Mr. Wells, but he does not go far enough: he stops short; he does not really admit a full-blown separate permanent individuality. He likened us to the Jungfrau or the Wetterhorn, which are only peaks in a range of mountains; and by the range of mountains he does not mean what I have been talking about—a larger self not yet fully realised, only partially incarnate in matter, not fully displayed by the temporary organism;—the range of mountains in his parable means the altitude at present attained by higher members of the race, and is to emphasise their connection with each other and with the whole. His individuals only exist for their brief span, in subservience to humanity in general; they arise, flourish, and decay; man, the whole human race, is the permanent enduring entity. So he claims.

But I would remind him, on astronomical grounds, that the human race cannot be immortal; its duration depends on the duration of the planet, and that must terminate, so that all the struggle and effort must, on his view, evaporate into nothingness:

> The cloud-capp'd towers, the gorgeous palaces,
> The solemn temples, the great globe itself—
> Yea, all which it inherits, shall dissolve;
> And, like this insubstantial pageant faded,
> Leave not a wrack behind.

So the anthropological Point of View is incomplete and unfinished. Life on the planet can exist for a long time, many million years perhaps, but sooner or later it must stop, and the whole long effort, all the striving and the aspiration, be wiped out, and leave no trace. This would follow from the incomplete truncated merely biological view, that the individual is nothing more than a material organism, and has nothing more than a racial permanence. Whereas the truth is that while each individual serves the race, the race itself is an abstraction, and is composed of individuals, each of whom has a permanent importance. The individual is more permanent than the race. My bodily self is but a small fraction of a whole, but, however insignificent in itself, it is an outward and visible sign of a permanent entity in the cosmos, something that endures and need not cease to be.

Mr. Wells wondered what was the essence of Christianity. Well, that is part of the essence, that the individual is the important thing. Each individual is of permanent and vital significance; the race is a congeries, a succession of individuals. Things do not exist in the abstract as generalisations; they exist as concrete individual entities. Reversing his well-sounding dictum, it is not man that is immortal, but men. It has taken a long course of evolution to produce a personality, and once produced it will continue. This is not speculation on my part, it is knowledge based on ascertained fact. Mr. Wells has a great deal of knowledge, but he has not got that; like other present-day amateur biologists he is probably not prepared to accept that, and therefore his Point of View stops short, frustrated, ending in nothing. So long as we contemplate matter alone, this is inevitable. Every structure made of matter is temporary. This world is temporary; it will cease.

But that is not the whole of existence: the permanent realities lie outside all that. They are accessible, not to the senses, but only to the mind, especially to the mind of the poet, the artist, the philosopher, and the saint. This is where the plain man has to that extent an advantage over the specialised scientist; he has primary apprehension of a number of most important things. Whatever he may be ignorant of, he is aware of his own consciousness; he can take his stand with Descartes and say: "I am sure of one thing—I can think." And he could go on to say: "All the complicated things that you tell me of, the vibrations, the condensations and rarefactions, the electromagnetic impulses, have to be presented to my mind for interpretation. Apart from my thought and instinctive apprehension they have no meaning. Neither colour nor tone exists in your scientific world; they are mere rates of vibration. It is through the action of the mind that they have any meaning, and you scientific people, that attend so closely to the machinery, may, if you are not careful, let the meaning evaporate and lose the æsthetic enjoyment altogether. You may let the soul of things escape you."

Yes, that is what the plain man might say. And by the plain man I mean the statesman, the man of business, the millionaire, and the squire. He knows nothing of the mechanism he uses: indeed, by himself he would not have any mechanism, except the one he inherited from his ancestors. He is dependent on the few who spend their lives in exploration and study. And the scientific worker might retort to the plain man thus: "You are not half grateful enough to us who undertake all the labour, and are content to leave the enjoyment to you. You are ready enough to take advantage of the machinery we provide, engines and dynamos and telegraphs and aeroplanes; you are even glad to have your own internal machinery overhauled and put right from time to time, but you seem willing to put up with any amount of pain and inconvenience rather than provide the seekers after knowledge with what used to be called 'the sinews of war', but which should rather be called 'the sinews of peace'. Give scientific men in general a reasonable opportunity, provide them even lavishly with what they need for their work, and a lot of avoidable evils will gradually become things of the past. It is not good sense to exult in ignorance: ignorance is not bliss, nor anything like it. You miss another kind of enjoyment through your scientific ignorance. We do not miss that. Nature is luminous to us; it surpasses every other work of art; it leads into the transcendent, the unknown. You say that we run a certain danger in concentrating on the mechanism through which reality appeals to our senses. We admit the danger. It is true that we are apt to get so entangled in the machinery that we ignore everything else. Some of us have foolishly denied that there is anything else. But we are becoming awake to the danger; we need not fall into the trap. We, too, have minds that can gain access to enduring realities; the soul of things need not escape us; we know that it lies in the unseen, the unsensed world. We, too, are able to recognise that as the world of reality."

This unseen or invisible world was referred to by Mr. J. B. S. Haldane in his remarkable address. He gave us something more than the biological view; he is himself a professed biologist, but appears to be a philosopher too. I have not yet been able to read his utterance, but when I heard it, it seemed to me sound and interesting. I can go nearly all the way with Mr. Haldane, at least in his present talk, but even he stops short. He does not admit the permanence of the individual. That frustration will sooner or later damage his system of philosophy, and constitutes its weak point. Actuality is always better than our schemes or ideas about it. In actual fact the individual really does continue, and no system of thought which does not include that great fact can be really satisfactory. That it is which in the long run gives meaning to the whole, and interprets the problems of existence. The Christian revelation also drives home the permanent and paramount importance of the individual soul. Ignore that, and our philosophy is defective.

Practical Outcome

Meanwhile, we can all agree about the importance of earth life, the life of humanity on the planet earth. We do not differ much about practical details, or about our ability to co-operate for the good of the whole race of man. The individual and the race must continue to evolve together; it is our privilege to serve the whole, and corporate humanity has a great future before it. If man takes the reins into his hands, and with the help of Higher Powers aims at concentrating on things of real value, no one can foretell what he may become. The earth is a great opportunity for development, and evolution has not ceased; man may rise in the scale to unexampled heights. That is what all the long struggle and effort in the past have been for. In no other way could man have risen from lowly beginnings, until by his own freewill he becomes something akin to the Divine. This is the deep meaning of earthly existence. This is the far-off event towards which the whole terrene creation moves. In moments of insight we recognise, with awe and devotion, the reality of an unseen, a spiritual world, the coming dawn of a glorified humanity, and the ultimate meaning of existence.

We have listened to great thinkers freely talking and brilliantly attempting a reasonable point of view. Let each listener think for himself, and be thankful for this great organisation which stimulates thought throughout the length and breadth of the land. May I leave two thoughts with you? "Desire earnestly the best gifts", and "We needs must love the highest when we see it".

NOTE.—A letter written by Sir Oliver Lodge, after Mr. Dickinson's Summing-up, constitutes a supplementary communication, and will be found on p. 145.

VI

SIR WALFORD DAVIES

FROM WHO'S WHO (1930)

DAVIES, SIR (HENRY) WALFORD, Kt., *cr.* 1922; Mus. Doc. (Camb.), 1898; F.R.C.M. (Lond.), 1926; Hon. LL.D. (Leeds), 1904; F.R.C.O. (Lond.), 1904; Hon. F.R.A.M. 1923; Hon. LL.D. (Glasg.), 1926; Organist of St. George's Chapel, Windsor, since 1927; Director of Music and Chairman of the National Council of Music, Univ. of Wales, since 1919; Gresham Prof. of Music since 1924; *b.* Oswestry, Shropshire, 6 Sep. 1869; *y. s.* of late John Whitridge Davies and Susan, *d.* of Thomas Gregory, Wem; *m.* 1924, Margaret, *o. c.* of Rev. William Evans, Rector of Narberth. *Educ.:* private tuition. Entered Choir of St. George's, Windsor, 1882; assistant organist to Sir Walter Parratt, 1885–90; also organist of Windsor Park Chapel Royal; held a scholarship (for Composition) at the Royal College of Music, Kensington, 1890–94; organist and choirmaster of St. Anne's, Soho, 1890–91; at Christ Church, Hampstead, 1891–98; Organist and Director of the Choir at the Temple Church, 1898–1923; Organising Director of Music, R.A.F., 1918–19; teacher of Counterpoint at the R.C.M. 1895; Prof. of Music at Univ. Coll. of Wales, Aberystwyth, 1919–26; Conductor of the Bach Choir, 1903–7; London Church Choir Association, 1901–13. *Published Compositions:* an oratorio, "The Temple" (Worcester Festival, 1902); a cantata, Everyman (Leeds Festival, 1904); A Sacred Symphony in F (Hereford Festival, 1906), a Choral Cycle called Noble Numbers (Hereford Festival, 1909); two other Song Cycles, Songs of a Day and The Long Journey; Five Sayings of Jesus (Worcester Festival, 1911); Song of St. Francis (Birmingham Festival, 1912); a cantata, Hervé Riel; Cycle of Songs for children, A Merry Heart; Fantasy for tenor solo, chorus, and orchestra from Dante (Worcester Festival, 1920); Choral Cycle, England's Pleasant Land; Cantata, Heaven's Gate (People's Palace Festival, 1917); Men and Angels, six Choral Songs with parts for tenor solo, chorus, and orchestra (Gloucester Festival, 1925); High Heaven's King (Worcester Festival), 1926; Parthenia Suite, Festival Overture and Holiday Tunes for Orchestra; a Children's Symphony (Hereford Festival, 1927); Quintet for Piano and Strings (Hereford Festival, 1927); Conversations for Piano and Orchestra, etc.; Peter Pan Suite for string quartet; church services and anthems; two sonatas for piano and violin; six pastorals for four voices, string quartet and piano; Browning's Prospice for baritone voice and string quartet; songs; Editor of the Fellowship Song Book; A Student's Hymnal, etc.

VI

SIR WALFORD DAVIES

IF the invention of wireless is going to empower men to meet once a week to deliberate as a world-court of listening minds with chosen speakers who come to the sacrifice, the jury will always automatically require that their witnesses shall at once be themselves and quite forget themselves, since they speak at the tribunal of the common mind of man. Men are their jury, but man is their judge. I will try to remember therefore to concentrate on things that matter to all men and avoid things that matter only to one.

I find that I shall probably be impelled to speak chiefly of love of life, of art, and also of emotion—that involuntary sign that the love of life is functioning. The old man who said "I likes music when it makes me feel creepy all down my back" expressed something familiar to all of us in one way or another. But he was apparently more intrigued with the symptom—emotion experienced—than with the love and loveliness which brought it about. That was natural, but surely a pity. For in life in general and art in particular, those men seem most trustworthy who put love and loveliness foremost, recognising these two as first causes of the experience called emotion, which they invariably put second, though it may affect them most at the moment. They do not consent to make emotion an end but take it for what it is—an essential symptom that something good has happened; a sign that the conscious need within us and the arrival of supply without, have conspired to a thrilling point.

Before we look into this a little closer, let me say that by love of life I mean literally love of everything lovable in life. To give examples: personally I love the sensation of Devonshire cream on my palate (not too much of it), and the scent of red roses; the touch of smooth cool metal when my hand is hot; the sight of sunset-red; and the mere sound of unison strings on the orchestra. Then, like all other people, the touch I love beyond anything is that of my beloved. I love doing things with set purpose; and still more if the doing is a definite exercise of natural skill; and this makes games lovely. Next I love wonderment. I love wondering about created things and about things never yet experienced, and still more I love trying to create music I have wondered about but never experienced. First and last, like other fellows, I love life, and find it all simply lovable except when I'm too busy bearing pain or vexation to be able to remember love and loveliness. I call this all by the common name, the love of God, since I see no other name for it: it is so lavish and unreasoned and unreasonable. I hope the above is a truthful, representative catalogue of the different kinds of lovable things. There is, you may remember, a far more vivid and moving list in a familiar poem of Rupert Brooke.

Searching for my Point of View, then, I find that all love of life (enthusiasm) and all consequent emotion (symptom of attained enthusiasm) are truly of two kinds to me, and two kinds only. I may call them creaturely and creative. Music, which I happen to love from childhood, is only conceivable to me as definitely and simultaneously a two-fold exercise and experience, a pleasure of the senses and a joy of the mind. Further, I find that my two kinds of joy seem to run, not only together but, at present united inseparably, though in an infinite diversity of ratios. What is true of music seems true in everything I feel or do in this life. I seem to observe it in my friends, too, especially in boys. Men seem from boyhood incurably devoted to two orders of delight and interest, which I can nearly always track down to creaturely pleasure of sense, or creative and contemplative joy of the imaginative mind. When first a baby says, "I want to see the wheels go wound", I believe he has begun his new and lasting joys, which cows and dogs, so far as we know, never begin (though a systematic game of a dog with a stick looks suspiciously like a beginning). So, personally, I have long been baffled and rebellious at what has been called by Hadow the "ordinary tripartite division of human nature which has passed current ever since the time of Aristotle". I seem to need and discern a different explanation of my own experiences definitely bipartite (hideous word!).

But lately it has become more possible to understand the enormous stress laid upon emotional experience as an actual aim, especially in art, because I seem now to see that what we call emotion (though it is not a faculty I exercise, but something that happens to me, like sunshine) is the essential symptom of the arrival of my desired joy, not to be mistaken for the joy itself. Emotion is the sign that a composite longing within me has been met and unified by a corresponding supply of experience without. And this seems a crucial distinction. If I did not make this distinction I should become an emotion-monger. To seek emotion as an aim I find is bad for me, and I find no inward sanction for such a search. But to experience emotion unsought seems altogether splendidly good for me; it not only comes as a sanctioned result or symptom of my other two natural searches—be they at the moment what is known or thought of by others as naughty or good (God and my own conscience alone bandy those two words about)—but its arrival gives me driving power which conscience bids me turn into act. Now art I understand to be act under steam of given emotion. Emotion as an end sought and enjoyed for itself seems damnable waste to me. I feel guilty whenever I consent to it. On the other hand, emotion turning to new act which in its turn brings new emotion has profound sanction, and I'm (literally) blessed if ever I do it, and so, I believe, are you and the others.

I shall now try to imagine that I am personally addressing my Judge—that Man of eternal commonsense and common spirit on whom Mr. Wells and I rely:

> "My Lord Judge, I want to ask whether there really is in music the element which makes it one of your Lordship's lasting interests. It is clear that our mortal bodies can happily and innocently jazz to it for a few years, but am I justified in the belief which I cherish that your undying Lordship is also concerned with it? Is it true that music is an important human index in that it is distinctly and always both a bodily and a spiritual act? A good swinging rhythmic tune undoubtedly can refresh tired bodies. It can be a physical diversion and stimulus. But why have I also found that a beautifully balanced melody played quietly with the minimum of physical stimulus to a large promiscuous prison audience at Cardiff—such a melody as that of Gluck in F from the Elysian music in 'Orfeo'—secured rapt silence and the most piteously enthralled attention from people down, out, and found out?"

I find my Point of View is gradually becoming one which assumes that I myself am like all other men I have ever met. This assumption has grown after four years at the microphone into a strong working belief; though I confess that from some individuals who write to me all about themselves or about myself I feel awfully estranged, until they manage to write about something of common human interest, when I am completely released again into working faith and order. Further, I find this oneness proved most of all in our having all been born with these two contrasted and equally natural though by no means equal or equalised capacities: one for what may be called a childish delight in creaturely interests, and the other for what may be called a child-like delight in creative interests. Neither for the presence of physical pleasures, which I see to be passing, nor for the pleasures of the mind, which I suspect to be lasting, do I feel any responsibility. I did not induce them, I found them there. For their ordering I do feel conscious of responsibility. (And I agree with an atheist neighbour of mine who once declared that though he didn't believe in God, yet when he was happy he "did want someone to thank".)

I think the next important thing to speak of at the moment is the varying ratio of creaturely and creative enthusiasm in life as in art. The ratio of interest at birth is apparently 100 creaturely to 0 creative. Children remain with a heavy percentage of creaturely egoism, both clamant and often amusing. But parents watch with wistful interest the gradual change of the ratio. From quite early childhood to late old age there are infinitely diverse ratios, not only as between one man and another, but in the same man at different ages and in different surroundings. For example, the things that fill you with a glow and thrill of emotion when you are suffering pangs of physical hunger will not do so when you are replete with nourishment or physical energy. I have read that Arctic explorers, who had chosen an adventure of heroism and great privation, were yet moved to talk by the hour, day after day, of the first meal they would have when they got back. The very smell of that first meal will probably bring a thrill to these heroes. The dire need and creaturely enthusiasm at last to be met will bring an experience of well-being that will momentarily seem the sum total of God's goodness to them, and the result will be nothing less than an emotional thrill over a beef-steak. This is decided for them by their 75 per cent. physical desire and 25 per cent. desire for more enduring things. That would not prevent the reverse ratio obtaining under different conditions the very next day, when, deciding to embark on another heroic adventure involving as much privation and endurance as ever, they have perhaps a thrill due to 80 per cent. heroic and only

20 per cent. creaturely reasons. These are quite imagined guesses to suggest the probable fact that there is no sharp division between men. We are obviously not divided or divisible into saints and sinners, heroes and unheroes, gentlemen and cads. Heroes may, on occasion, get a memorable thrill from a sensational meal, and, conversely, gluttons may feel uplifted by heroism: in the common acceptance of the terms all men might be heroes, and all heroes are but men. So long as a spark of willing energy remains, men are conscious of this personal will-amalgam in them that is never 100 creative and 0 creaturely, for that would be superhuman, nor 100 creaturely and 0 creative, for that would be bestial. But it is good to remember Socrates and Stephen, who at the cruel deaths of their bodies gave to history serene and certain witness by words, by demeanour, and a countenance "as it were of an angel" that men can actually arrive at a ratio of joy which is nearly 99·9 per cent. spiritual. It then becomes less difficult for honest, hard-thinking, ordinary men to conceive that the very creative Spirit of the Universe did actually and completely enter once into one man upon this planet at one spot. The incredibility of this unique happening to natural man is lessened, if not removed, by Christ's own reported expectation that his followers would do "greater works" than he. If one man once maintained in the body, from the first

moment of awareness to his last breath, the spirit of wholly selfless enthusiasm, those who believe him very God and very Man will not seem too blameworthy. On the contrary, those who would say that Christ's ratio was anything less than 100 creative enthusiasm would, one imagines, have hard work to support what they say with reasons.

Here I seem at a point where I must mention my belief in immortality. It seems different in detail from that expressed by Mr. Wells. And I cannot with Sir Oliver say "I know", though I might say it with certitude, in Handel's way.

I believe in our oneness. But I do not believe that the price of our progress into unity is the loss of our diversity. It makes the same kind of nonsense to my mind to say that all Beethoven's Symphonies will only survive as one immortal symphony as to suggest that all good men—so charmingly and humorously diverse—will survive as one man. Granted it is hard to believe that the exact created entity called the Moonlight Sonata is imperishable, it is still harder to find anything perishable in it. All the pianos in the world, all the copies of the work, all the hands that ever played it, all the brains that stored it, are, it is true, perishable bagatelles, for you can burn the animal machine called the brain. But can you really burn or otherwise destroy an individual form? In the same way the lilies perish, but the form of a lily, as a creative concept of God and a percept of His appreciative little son called man, seems to have no perishable quality.

Mr. Wells carries us far with his sentence, "I think man is immortal"—an immense lead and help to the "Let us eat and drink for to-morrow we die" part of us. Granted, however, that lilies fade and men go to dust, is there any sign that in an infinite and presumably ever-expanding universe of energetic forms of measureless diversity there will be no room for the survival of the creative energetic formers? I would appeal to Mr. Wells to say whether he sees no signs of there being room in this exuberant universe for the survival not only of every form of beauty once conceived, but for the conceiving workmen, too. Is the survival inconceivable of an infinite number of creative children of God who can be one both in Mr. Wells' sense and in Christ's sense?

So now, coming as a Point-of-Viewer to your Lordship's homely tribunal and, within the hearing of fellow-jurymen, called upon to give the sum of my Point of View, I find that four great things certainly exist in the universe, and that I love them and believe them to be lasting.

(1) Energy is; I love Energy, when I can bear it, and I believe Energy eternal. (2) Form is; I love it endlessly, when I can comprehend it, and think it eternal. (3) Movement is; I love Movement and see no end to it in the universe. (4) Recurrent Change is; when I can fit myself to it, I never tire of it, and I believe Change and Recurrence—like the spontaneous forces which mould a well-tempered Rondo in music—are inherent in creation and eternal. The galactic universe is perhaps one stupendous Rondo of the Almighty.

Music seems one of the practical expressions in ordinary life of our natural devotion to these four exhilarating immutables. Indeed, it seems itself pre-eminently a language of energy, form, movement, and change imaginatively wielded, purposefully blended. It is true that we may find semblances of their opposites manifest in music as in life; apparent absence of strength, apparent chaos or formlessness, dullness, stillness, negation, death. But they are all naturally abhorrent and only believable as appearances.

May it not be that the enthusiasm we sum up in the name "love" is the creative energy that can fling itself into chosen form, both infinite and infinitesimal, with ceaseless movings and unchanging changelessness.

And here I would venture to add a question: Why do so many of to-day's most honest, capable, distinguished, courageous thinkers allow themselves to leave Christ out as a thinker and an influence upon thought? As Liddon long ago said, so many men leave the university ready to exclaim, "I live, yet not I, but Plato liveth in me". But what has come to pass that Christ's commanding philosophy of creative obedience and creaturely endurance should be unmet or even unmentioned in the procession of human thought in A.D. 1929? Would that Plato himself could give a post-Christian pronouncement at the microphone.

All men I have ever known at all intimately seem in their own way to love and pursue, to watch, listen to or otherwise seek to partake in such doings as seem to offer a full cup of energy, mastery, and wonderment. Over and over again when I am thrilled by any act or art—a piece of music, a book, or a remark—somebody quite different from myself will say, delightedly, "Oh, did you like that? I was thrilled, too". This is heartening. And surely it means that those who, like composers of music, are impelled to throw their enthusiasms into audible forms have no escape from certain common creative and neighbourly obligations. Prophets, seers, poets, all authors alike, down all the ages, must deal in enjoyabilities, intelligibilities, and what may be called wonderabilities. To deal in intelligibilities without quickening wonderment is to become prosaic. Conversely, to deal in wonders without intelligibility is to become "idiotes". It is comparatively easy to say old things in heard-of ways or to say a brand-new thing in an unheard-of way. But to say a new thing in a familiar way at once quickens and satisfies the hearer. In my view, the later music of Schönberg and much other contemporary effort is new enough to be wonderable. But how much more wonderful it might be to-day and at once if it were both enjoyable and intelligible as well! The fact remains that it needs the grace of Heaven, which we call genius, to make the world's next remark with swift and perfect relevance.

VII

A SUMMING-UP

VII

A SUMMING-UP

G. LOWES DICKINSON

SINCE it falls to me to give the last talk of this series, I will begin with one brief observation. We did not choose ourselves, we were chosen by the B.B.C., and the B.B.C. did not choose us because of the character of our opinions. If we agree in certain points, we disagree in others; and we lay our views before you to consider them with such approval or annoyance, interest or boredom, as you may happen to feel. Another set of people might give quite a different picture, and for my part I hope they will; especially I should like to hear what the young are thinking, for they are sure not to be thinking the same as any of us. In admitting these controversial talks the B.B.C. is doing a great work of education, and those who direct it are well aware that every kind of opinion must have its chance of expression.

I am not going to enter here into controversy with the other speakers. What I propose to do is to point out where we agree or disagree, and to add some observations of my own on points which I touched lightly or not at all before.

I will first say a few words on the political topics

which have taken, perhaps, too large a place in our discussion. We all of us, as it happens, are democrats in some sense, though we all quarrelled with the others' use of the word. However, it does not matter about the word if we are agreed about the thing. I used the word to mean merely government by consent, not by force; and so far the other speakers agree. You may think—I shall be delighted if you do—that that is a commonplace. It may be so in England, but it certainly is not over a large part of Europe, not to mention Asia, and perhaps it may be questioned, in practice, in England sooner than we suspect. Then will be the time to show what we really do believe.

The other point on which we all, except Dean Inge, seem to agree is on the need of drastic changes in the present system of property. That, of course, is a point far more controversial than the other, but it is essential if we are really intending to have anything like equal opportunity; and there I will leave that matter. It will certainly be a burning question during the lifetime of the young and the middle-aged, and they will have to take sides about it.

Lastly, we are all agreed about the folly and waste of war, and this, I believe, is a pretty general feeling now in this country. It needs translation into action, and I believe, as I said before, that the League of Nations is the only practical means of doing that. I will not now dwell further on the matter.

Turning now away from political questions, I will speak about our attitude to science. On that point I suppose Mr. Shaw [1] disagrees with the rest of us, and I am inclined to congratulate him upon that, for he is never happier than when he is disagreeing. The rest of us seem to be agreed about the place of science. Dean Inge delights me by saying that "the air that blows round science is like the air on mountain tops, cold and thin but pure and bracing," and he adds that "to seek for the truth for the sake of knowing the truth is one of the noblest objects that a man can live for." That, I believe, we all endorse. I agree also with the Dean when he says that scientific men have constructed very poor philosophies; if, as I suppose, he means that men of science have been and still are very bad scientists on subjects lying outside their speciality. Where they are not personally investigating they fall back as easily on prejudice as any men not scientific. But that is because, in the regions in question, they are not acting as men of science but as men in the street. There is, for instance, no presumption at all that physicists will be more intelligent than other men about politics, or psychical research, or values, or religion.

Possibly, however, when the Dean says that men of science make bad philosophers he implies that there is some other method than that of science for

[1] See note page 21.

reaching truth about matters of fact. To that I should not myself agree, nor I suppose does Mr. Haldane; for he says that the churches are half empty to-day because their creeds are full of obsolete science. Nor, I think, would Sir Oliver Lodge agree. He does not speak of two methods of knowledge. The point on which he is at issue with Mr. Wells and Mr. Haldane is as to what has, in fact, been discovered by science. He believes that individual survival of death, or, more simply, personal immortality, has been proved. But he believes it on evidence which can be tested; and presumably if a consensus of intelligent and honest scientists should show that the evidence is inadequate, or leads to some other kind of conclusion, he would admit that his own conclusion ought to be abandoned. I do not propose here to go into the character of the evidence: that is a special subject of singular complexity which most men of science have not even begun to consider at all.[1] There is, however, another point in this connection which I would like to make. Sir Oliver regards survival as a fact that increases the value of human life. I would say that it depends on the kind of life into which we survive. Nothing depresses me more than many of the accounts purporting to proceed from the dead about their occupations on the other side. They indicate a continuance of what we have most of

[1] See the letter by Sir Oliver Lodge, printed on page 145.

us had quite enough of here, houses, pianos, lawn tennis, whiskies and sodas—in short, life as we know it with all the disagreeable facts ruled out; and to this I confess, if I know myself, I should prefer extinction. Nor, to tell the truth, does the picture given in much religious writing appeal to me. There used to be a story told of Mr. F. W. H. Myers, a man of genius who devoted his whole life to investigating this question, to the effect that he was once pressing his host at dinner as to what he really believed or hoped about life beyond death. Driven at last into a corner and forced to reply, the unfortunate man exclaimed: "Of course, if you press me, I believe that we shall all enter into eternal bliss; but I wish you would not talk about such disagreeable subjects." It has always seemed to me that the words, apocryphal or not, represent the real feelings of many ordinary men about such accounts of the future as are contained in many hymns and sermons.

The great poets come nearer to depicting something one could really desire. Dante, for example, has reached further than one might think possible to language in his description of Paradise. But his account comprises also Hell and (what seems to me more probable) Purgatory; the relegation of pagans, however good, to a limbo on the very verge of hell; and much else which has become, or is becoming, incredible to modern men. Sir Oliver himself, wisely I think, does not give any description of the life he believes to follow life on earth. Whether that life exists, and of what character it is, science might conceivably one day determine beyond reasonable dispute, and then much of our present discussions will be out of date; but, as things are now, I do not know whether I want to be immortal or not, for I do not know what kind of immortality it will be. I only know there are many kinds I should hate.

Meantime, as long as our ignorance remains, I think the wisest course to take is one recommended by Goethe. "A sensible man", he says, "who wants to be decent here and so has to struggle, fight, and work, leaves the future world in peace and is active and useful in this one". But leaving it in peace does not mean denying it; it means—and especially it meant to Goethe himself—going on, with that horizon and all horizons open, and capable of being filled, not with dogmas, but with anticipations clad in the forms of poetry and art.

And this leads me on to Mr. Wells' talk in this series, which I read with great interest and sympathy; for my own thoughts for many years have led me in the same direction. Whatever may happen after death there is, in my belief, no doubt about the course that life can take and should take here. It is the course Mr. Wells describes with his usual skill and sincerity. What he says is, for a great part, the undoubted experience of us all. Much of us all, as we grow older, passes away and perishes. Also much enters into us from sources outside ourselves, from the talk of our contemporaries, from the writings of men long dead, from the whole history of our society and of the society of the whole world. Also our character is largely determined by our ancestors. We are not, either biologically or in any other way, isolated individuals. We are parts of a whole, and a great part of our education is learning to realise this fact. I need not repeat Mr. Wells' illuminating exposition, but I do not think there is anything in it with which I should disagree.

The point of controversy seems to come in on the question of individuality. The best of us here, Mr. Wells says, are not getting *into* individuality, but getting *out of* it; and it is our proper business to do that so far as we can, using our own special gifts, "for the good of the master-being Man". This getting out of the limitations of oneself, as well as one can, seems to me to have been the result, if not the object, of the lives of the men I admire. They did it commonly by dint of very painful efforts and very hard knocks. For we all begin as ignorant, greedy, egoistic creatures, and perhaps great men do so more, rather than less, than others, for they have more vitality and therefore more capacity for evil as well as for good. The question, however, arises: What is the character of that into which they grow? Mr. Wells says it is the immortal soul of the race. Sir Oliver Lodge demurs. The race, he says, is an abstraction, composed of individuals each of whom has a permanent importance. Clearly, when he says this, he does not mean the individual Mr. Wells wants to get rid of. What then does he mean? The real individual, he says, is only partly and superficially expressed in the individual as we men know him. He used the analogy of an iceberg, the greater part of which is buried beneath the sea, and of which the submerged parts are the important and enduring ones. The recent discoveries of psychology show that below the conscious individual there is a more extensive sub-conscious one. But it seems to be as yet an open question whether that region belongs to the individual or to something more universal. Certainly it is full, among other things, of very primitive and by no means admirable things; but also of very sublime things, if it be true, as Mr. Myers used to say, that genius is an up-rush from it, invading the ordinary consciousness. I should, on the face of it, be inclined to believe that we are indeed submerged, as Sir Oliver says, but that below the surface we are all connected. I should use the metaphor of a concealed reef of rock, emerging, at points, into what look, on the surface, like detached islands. The facts of the subconscious would seem to suggest that, at least as much as the other. Meantime, however, the important thing is this enlarging of the individuality into some calmer, more detached and more disinterested being. That we know to be possible to men because we know men who have done it, and are doing it, at any rate in part. And that, as it seems to me, is the thing the best men always desire to do, when they get freedom enough from economic stress to do what they want. I agree with Mr. Wells that the whole machinery of democracy or Socialism is valuable mainly because it may lead to that result.

We may perhaps get a little further in making clear to ourselves what this attitude is, which great men partly achieve and smaller men attempt. It is, in one aspect, science, conceived as Mr. Wells and all great men of science conceive it—a progressive and tentative knowledge, disinterested in its purpose, sincere in its method, imaginative in its outlook. In another aspect it is art, including in art all good poetry and literature; for this has always the quality of release, of raising the consciousness to a higher and larger plane. Art does not argue nor agitate nor preach. It lifts. And if it were possible to experience real life as a great poet or dramatist or novelist represents it, we should have got further than in fact we can, made and placed as we are. As it is, we oscillate between brute fact and the representation of it in art. But that also is part of our straining up towards that higher consciousness.

I will add, in this connection, a word on a subject which has hardly been mentioned in our talks, and which does not lend itself well to public discussion, though none is more convassed in private talk. I mean personal relations. No theme is more frequent in literature and art than the theme of love, treated in every possible way, passionate, satirical, bitter, ironic; for experience leads to all those attitudes. The Greeks used to distinguish a vulgar and a heavenly love; and there is, in fact, every range, from crude lust to the love of the divine, and every mixture in between. Great poets have commonly regarded love as one of the ways by which the self can be raised to a higher level. Dante symbolises it in Beatrice, the girl he worshipped in his youth, and Goethe in the Gretchen of his *Faust*; and a philosopher who died the other day made it the centre of his whole view of the universe.[1]

Love in that extremer sense is perhaps rather an ideal than a fact, for it seems to assume that the conflict between sense and spirit has ceased. But in some form or other this particular relation is certain to preoccupy most people far more than either science or art. The young now, so far as I know them, are franker and freer, both in their talk and their actions, than they were when I was young myself; and that is all to the good. Few things could be worse than the taboos of the Victorian days and the ignorance, confusion, and hypocrisy which accompanied them. Modern men and women are com-

[1] Dr. J. E. McTaggart.

paratively free. Where there is freedom there is, of course, always the chance of going to the Devil; but without that chance we have no chance of getting to the other place.

It will be seen that, while I agree with most of the other speakers in this debate that science is the only way of attaining knowledge, like them I do not suppose knowledge to be the only avenue to the higher life. The urge that is behind all life comes pressing in upon us through many channels, invading us often against our wills, driving and abandoning, creating and destroying, very great, very terrible, but never to be escaped. The best account of it I think is the one given by Goethe in the song of the Earth-Spirit:

In the storms of action, the floods of life,
I surge and sway
Above and below,
Hither, thither, to and fro.
Birth and death, an infinite sea,
A web that changes eternally,
A living fire!
I work at the loom of Time, I smite with the weaver's rod,
In the whirr and the roar I fashion the living garment of God!

And as I have mentioned Goethe, I will devote the few minutes left me to saying something about him, for there is no better example of the kind of development I am thinking of. Goethe was a man of action, a poet and artist, a critic and a scientist. He was a great poet—one of the greatest, I should say. He tried to be a painter, and I have been told by a modern critic that if he had lived now his gift would have been recognised as it was not in his own time. He was a scientist and spent more of his life on science than on anything else; but his science has grown out of date, as all science must, if it continues to grow. His administrative work, in which he spent ten of his best years, has of course long ago vanished. His significance lies in the fact that, more than anyone I can think of in the last few centuries, he was continuously possessed by the urge of which we have been speaking. He was no saint; saints are made of thinner, if purer, stuff. He was no perfect poet, for he was doing so many other things. He seemed to himself to have been constantly pursuing false roads; but he never ceased to pursue and in pursuing to grow. So long as man strives, says God to Mephistopheles, he will err, but through error he will develop. Our business, Goethe says, almost in the words of Mr. Wells, is to make ourselves immortal:

> Waste not a word
> On the things that must pass.
> To grow immortal
> That is our task.

And he found the task hard enough. "There is no sin," he says somewhere, "which I cannot conceive myself committing". "The sins of youth", he wrote, when he was growing old, "are under your feet, now you've the sins of old age to meet". At twenty-two he wrote the story of *Werther*, and fifty years later he said: "The man who wrote *Werther* at twenty-two finds seventy-two uncommonly tough." But through all this, the words traditionally said to be his last, "Light, more light", were the guide of his life.

Of that life the key was action in a broad sense. Action, of course, meant to Goethe action of every kind, practical, scientific, and artistic. It meant also love; and here too he had every experience, good, bad, and indifferent. He knew lust, he knew marriage, he knew the love of God. "Love", he says in one of his letters, "is at once the simplest thing in the world and the most capricious, in that capricious medley which we call man. Like a star, the path of which we can trust as we trust the course of the sun, and like a meteor that deceives the watcher worse than any will-o'-the-wisp". He had every experience of love, but he kept hold of its best and highest form. "Dislike and hatred", he said, "fetter the observer even if he has insight: they limit him to the surface. But let sympathy and love be married to insight, and then the world and mankind will lie open before him. He may enter on the path which will lead him to the highest height of all".

It is from this point of view that I would approach once more the question of religion. What we know certainly is this drive, this urge, however feebly we follow it. When we try to formulate its object in creeds we are certain to go wrong, because we are flinging words at what we only dimly apprehend. And what horrors have not been inflicted on the world by the attempt of men to force upon others their own imperfect conceptions! Men always want to fight when they ought to be trying to understand. The spirit of war extends into everything, into love and art and religion, as much as into politics; and it must be exorcised from all if we are ever to escape to a higher phase of life. Whether we shall do so or not I do not prophesy. I am neither an optimist nor a pessimist. But our future development may depend a great deal on what happens in the next ten or twenty years, and on the attitude of those who are now young. From what I know of them I believe in them; and if I have been quoting the words of a man who has been dead a hundred years, it is because I believe that he still lives by the virtue of his indomitable spirit. May I then quote, especially for the young, in halting English, the great song Goethe wrote for the Freemasons—called by Carlyle "the marching-music of mankind":

> The Mason wanders
> Like all who live.
> And all the wonders
> He strives to build
> Are like man's striving.

The Future will waken
New joy, new grief.
We see but one step,
And that step brief,
Yet press through, unshaken.

Dark in the distance
There hangs a veil,
Solemnly. Silent,
The stars are above us,
Beneath us the graves.

Behold them, and ponder!
Strange visions will rise,
Changing, baffling
The bravest eyes,
Dread visions, grave-hearted.

But listen! The voices
Call from the sky,
The Spirits, the Masters,
Unceasingly:
"Loiter not! Work!

"Here are woven,
Here in the calm,
For all who labour
Full crowns of palm.
We say to you, 'Hope!'"

VIII

LETTER BY SIR OLIVER LODGE

To the Editor of "The Listener"

DEAR SIR,—First I want to praise an article in the B.B.C. Yearbook: the one by Professor Eddington on the ether. Every word is significant, and it is full of telling illustrations. I go further in some respects, but I agree with this article profoundly.

Next I want to make a slight additional comment on the "Points of View" series: especially on Mr. Lowes Dickinson's summing-up. So far I have only heard it, but I suggest that he has rather spoiled his mainly serious and understanding treatment by a misconception concerning some testimony supposed to be from "the other side" about conditions of future existence. Giving the genuineness of such testimony the benefit of the doubt, we may admit that it cannot be easy to describe things to people like us with no experience of them: the only plan is to indicate their likeness to things already known. One piece of information to be conveyed is, let us say, that music does not cease with earth life, that the essential reality even of instrumental music persists: in that sense the term "harp" has been used from early times, and sounds dignified, while a modern term like "piano" sounds frivolous.

"Piano" is therefore introduced to indicate contempt. The device is often employed; it seems quite effective before breaking a treaty to call it "a scrap of paper." The quotation "whisky" is often misused, not as illustrating the inappropriate things that might be called for by young soldiers shot over in droves during an epoch of slaughter in the Great War—a transition so sudden that they do not realise what has happened and think they must go on fighting—not that, which may be true, and anyhow is reasonable, but as if alcohol had been represented as one of the normal ingredients of life there: which would be absurd. "Spend their time in smoking and drinking" was what one clerical opponent seemed to think a fair misrepresentation. Well, it is one way of showing contempt; but men of standing and education ought to know better.

A curious argument, which many people seem to think it appropriate to use, is that certain things cannot be true because they wouldn't like them. The major premise of this argument is, apparently, that the plan of the universe must be such as to meet the approbation of any prejudiced and uninformed mind, who in default of such approbation would "shatter it to bits and then remould it nearer to the heart's desire".

That reality is always better than our conception of it, not worse, I admit; but that we should be competent to reject anything on grounds of mere dislike, in face of positive evidence in its favour, is surely extremely doubtful. People have disliked all manner of things that they only half understood, and that have turned out true. Some people still dislike the idea of evolution, and just hate the animal ancestry of man. At one time fossils on hilltops were disliked, unless they were associated with the story of Noah. Professors and clerics deeply resented the notion that the earth was a ball careering through space, and threatened to torment or even burn anyone who promulgated such a preposterous doctrine. Nowadays people are content merely to ridicule those engaged in psychical investigation; though they still hold the threat of fines and imprisonment over the unfortunate instruments who are used in the inquiry. No end of things can be negatived and rejected, if superficial likes and dislikes are to be recognised as arguments. Learned people still fall into the old traps.

Everyone seems liable to fall into some trap or other. Who shall venture to claim that he is exempt? I do not; but I do like to point out the beam in other people's eyes, even if I fail to recognise the mote in my own. No one need take the trouble to point out that this is upside down. I know it, and am ready to give myself away in every petty detail, in everything except when I know that I am on the track of truth.

One reason for dislike in continued existence is probably suggested by Mr. Wells when he expresses the ancient desire to escape from the burden of self. He has fine authority on his side, and it is somewhat of a confession to say that I am not conscious of that desire to escape. I should feel lonely if my old self, such as it is, were obliterated, or the memory of it disallowed.

Lastly, in a recent issue of a paper called *John o' London's Weekly* (November 16), Mr. Robert Lynd has written an admirable open letter to Mr. Bernard Shaw, with which I extensively agree. In it occurs the following sentence, upholding the desirability and usefulness of a sense of humour:

"It is a good thing that democracy—like romantic love, military glory, medical science, and a thousand other things—should come under the searchlight of the comic sense and prove whether or not it can come out of the test without loss."

Yes, and if psychical investigators are such feeble folk that they cannot stand up under the threats of antiquated Acts of Parliament and the gibes of contemporary humorists, they are not worthy of their vocation and might be better dead.

Mr. Wells said there was one thing we should never know: "We shall never know that we are dead." Well, there he is mistaken: this is just one of the things that in time we shall all know. And very likely we shall want to tell other folk what it feels like. Judging by present-day experience they will turn away and feel it derogatory to permit themselves to listen. Even over there we must bide our time and be patient.

OLIVER LODGE

November 20, 1929

GEORGE ALLEN & UNWIN LTD
London: 40 Museum Street, W.C.1
Cape Town: 73 St. George's Street
Sydney, N.S.W.: Wynyard Square
Auckland, N.Z.: 41 Albert Street
Toronto, 77 Wellington Street, West